DATA VIRTUALIZATION

Going Beyond Traditional
Data Integration to Achieve
Business Agility

JUDITH R. DAVIS AND ROBERT EVE

To Mary, Elizabeth and Katherine,
who delight and inspire me daily

– RJE

To Herb, who shared the journey

– JRD

Table of Contents

Foreword

Providing business users with the data they need to make effective decisions has always been difficult. Increasing data volumes, varieties of data and heterogeneous data stores are not making this task any easier. For the past two decades, the solution to accessing disparate data has been to consolidate the data into a data warehouse, and provide users with tools to access and analyze this consolidated data.

However, growing demand to access data not managed by the data warehouse, coupled with the need to access current data for more agile decision making, means organizations have to change the way data is accessed and used by business users. The solution is to give users a single interface to the data and data services they need, regardless of where the data is stored or how it is organized. This doesn't mean organizations have to abandon the data warehouse, but rather they need to extend it. Data virtualization is a set of techniques and technologies that provide the key underpinnings that enable organizations to do this.

Data virtualization is not new. It has been around for many years in a number of different guises, including enterprise information integration and data federation. It is only recently, however, that this approach has gained the recognition and success it deserves.

There are several reasons why data virtualization has taken so long to gain traction. Perhaps one of the biggest is that data virtualization was marketed initially as a way to build a "virtual" data warehouse and to avoid building a real, physical one. This approach was immediately assailed by data warehouse advocates as unusable and unworkable, and it took some time for data virtualization to recover from this assault and poor positioning. Like any technology, it is important to understand where data virtualization can be used, and where it can't. This is one of the key objectives of this book.

Another reason for the slow acceptance of data virtualization is that it was seen initially only as a technique for use in data warehousing. As the case studies in this book demonstrate, it has value for a wide range of enterprise data integration projects. This broad use is enabled not only by data virtualization's data integration capabilities, but also by its ability to act as a bridge to data supplied by middleware and services-based solutions.

For me, the feature of data virtualization that frequently goes unnoticed lies in its ability to provide IT and business users with a single high-level view of data that may be spread throughout the enterprise. This capability can dramatically simplify access to data for less experienced users.

Data virtualization is now a mature approach that has a wide range of uses, and what excites me about this book is the rich set of real-world case studies it contains. The proof of success of any technique or technology lies in its adoption, and these case studies will provide you with detailed information about how a wide range of different companies in a number of industries are achieving significant business benefits from the use of data virtualization.

This book should be read not only by IT professionals who are interested in, or considering the use of data virtualization, but also by executives and business users who are looking for ways of improving access to data for business decision making.

Colin White
BI Research
Ashland, Oregon
September 2011

About the Authors

Judith R. Davis is an independent analyst with over 30 years of experience in the business application of information technology. Her current focus is business intelligence and related products and technologies to optimize decision making across the enterprise. In her work with the Business Intelligence Network, Winter Corporation, DataBase Associates International and the Patricia Seybold Group, Judy has consulted with major vendors and end users and authored many in-depth research studies, white papers and articles for industry media organizations and technology providers. She holds an AB in economics from Duke University.

Robert Eve is the Director of Marketing for the Data Virtualization Business Unit at Cisco Systems. Prior to Cisco's acquisition of Composite Software, Robert was the Executive Vice President of Marketing at Composite Software, a leading data virtualization software provider. Bob has helped define the data virtualization category and authored nearly one hundred articles on data virtualization since joining Composite in 2006. His experience includes executive level marketing and business development roles at major enterprise software companies including Mercury Interactive, PeopleSoft and Oracle. Bob holds an MS in Management from the Massachusetts Institute of Technology and a BS in Business Administration from the University of California at Berkeley.

Acknowledgements

This book is the result of innovation, the kind of innovation that fuels superior business performance.

Originally sourced from advanced research on query optimization in the early 2000's, data virtualization has emerged in less than ten years as a significant data integration approach that is now widely adopted by large enterprises and government agencies.

This book, the first ever written on the subject of data virtualization, is dedicated to the pioneers of data virtualization. These pioneers include the enterprise software vendors that nurtured nascent research concepts and evolved them into complete, enterprise-scale product suites to offer customers a new way to integrate enterprise data.

These pioneers also include the information technology analysts and media who provided high-level understanding of user requirements, guidance and feedback on solutions strategy and communication of data virtualization's value to a broad business and IT audience.

Finally, and perhaps most importantly, these pioneers include the innovative people and enterprises that successfully adopted data virtualization and achieved superior business agility and performance.

The authors would especially like to thank the ten case study contributors whose data virtualization solutions are described in this book: Leonard Hardy, Michael Linhares, Mark Morgan, Craig Richards, Kenny Sargent, and Emile Werr, plus several others whose corporate policy does not allow attribution. These special individuals are all leaders of successful data virtualization implementations within their organizations, and have been graciously willing to share their experiences with others. Without their hard work and success, this book would not exist.

The willingness to contribute to this collection of data virtualization stories is validation of each organization's successful collaboration with Cisco. As such we would like to acknowledge the work of Cisco's engineers, field staff and administrative team. Without their involvement, these implementations would not have occurred. In addition, we would like to commend the collaborative efforts of the Cisco Customer Value team. Led by Bob Reary and supported by Juli Ring, this unique-in-the-software-industry team built the ongoing relationships with the contributors that encouraged them to tell their implementation stories.

Further, Mr. Eve would like to thank Jim Green, General Manager, Data Virtualization Business Unit at Cisco (formerly Chairman and CEO of Cisco), for his leadership and guidance and to acknowledge the excellent support of Cisco colleagues and friends David Besemer, Gary Damiano and Peter Tran, as well as Scott Humphrey of Humphrey Strategic Communications.

Excellence is also a quality we associate with the creative team at Westminster Promotions and Nine Five One Press including Marianne Lee, Lydia Klem, Amanda Ray, Jack Tse and Debra Valdez; and with Sangita Patel of Cisco (formerly Composite) who, as part of the creative team, coordinated execution of the myriad activities required to publish this book.

The authors would also like to thank Colin White of BI Research, another pioneer in business intelligence and data integration and a long-time friend, who graciously provided the foreword for this book. In addition, we appreciate the kind back cover thoughts provided by the wise and ever-encouraging Merv Adrian, Wayne Eckerson, Mike Ferguson, Claudia Imhoff, Shawn Rogers and Rick van der Lans.

And finally, the authors would like to thank each other for the great teamwork that has made our collaboration a wonderfully positive and fruitful experience.

Introduction

Business Agility Is About Survival

"In the struggle for survival, the fittest win out at the expense of their rivals because they succeed in adapting themselves best to their environment."

This quote from Charles Darwin's *Origin of Species*[1] in 1859 captures the essence of every organization's need to continuously adapt in response to a complex and dynamic business environment. To survive and be competitive, the organization must be adaptable. And adaptability requires the agility to quickly take advantage of new or changing business opportunities. Business agility, perhaps above all else, has become the most important overall success factor for every enterprise today.

Information Technology and the Pursuit of Business Agility

While the importance of business agility is well understood, achieving it is a difficult and ongoing challenge. The key to success is *information*. Armed with the right information, business decision makers can better evaluate their environment and decide

[1] *On the Origin of Species by Means of Natural Selection, or the Preservation of Favoured Races in the Struggle for Life, Charles Darwin, November, 1859*

how to adapt it for future success. Providing this information and the supporting infrastructure that delivers it to business decision makers is the responsibility of information technology (IT) organizations. In the quest to become an agile business, business and IT leaders must address all three elements of business agility:

- Business decision agility – The first is *business decision* agility. What happened in the past? What is happening now? What is going to happen in the future? Business leaders need the answers to these questions in order to identify and execute the ongoing business changes required for long-term success. This knowledge and insight can only be derived from complete, high-quality and actionable information. Without this information, business leaders must fly blind.

- Time-to-solution agility – The second element is *time-to-solution* agility. When the business seeks to enter a new market, launch a new offering, implement a new business process or radically shift an existing one, IT must quickly develop and deliver the tools and information necessary to support the analysis and decision-making process as well as provide any IT-based solutions that the new initiative may require. Failure to do so in a timely manner delays success, opening the door for competitors to capitalize on new opportunities instead. Thus, rapid time to solution is critical to achieving business agility.

- Resource agility – The third and final element is *resource* agility. In many enterprises, IT operations are one of the largest ongoing resource expenses and IT infrastructure is one of the largest capital expenditures. Whether the financial strategy is "more from more," "more from the same" or "more from less," maximizing the business impact of and return on these significant IT investments is crucial for long-term survival. Achieving the "same from the same" or worse, "less from the same," are not valid options in today's fast-changing business and IT environment.

On these points, CEOs and CIOs agree. According to IBM's 2011 global CIO study[2], based on face-to-face conversations with over 3,000 CEOs and CIOs, the number one priority for the next three to five years is the ability to derive insight from the huge volumes of data being amassed across the enterprise, and to turn those insights into competitive advantage with tangible business benefits.

Information Technology Challenges Abound

Effectively leveraging information assets to achieve business agility is complicated by the fact that the volume of information is expanding at unprecedented rates. An IDC research study[3] found that, in 2010, the universe of information exceeded one zettabyte, or one trillion gigabytes, for the first time in history, and is expected to grow another nine fold over the next five years.

These massive volumes are inexorably intertwined with the multiple generations of information technology that organizations have adopted over the years – mainframe, client-server, Internet and cloud. Simply maintaining this complex technology infrastructure, even at status quo service levels, strains most IT organizations today. This leaves IT with little opportunity and few resources to optimize this complex infrastructure for business advantage.

> *IT organizations face the daunting task of responding to constantly changing business demands ... Has IT's mission to support the business become an impossible quest?*

IT organizations also face the daunting task of responding to constantly changing business demands, many of which require timely development of new or revised IT solutions. Mergers and acquisitions are just one example. Another is the fact that supply chains must form and re-form seemingly overnight as product lifecycles shorten and products become more personalized. Further, with the explosion of social media and mobile computing, end users are adding new and unforeseen demands for fast access to information. Has IT's mission to support the business become an impossible quest?

---

[2] *The Essential CIO: Insights from the Global Chief Information Officer Study*, IBM Institute for Business Value, © IBM Corporation, May, 2011
[3] *Extracting Value From Chaos*, © IDC, June, 2011

In true Darwinian fashion, the business side of the organization is taking greater responsibility for fulfilling its own information needs rather than depending solely on already-burdened IT resources. For example, in a 2011 survey of over 625 business and IT professionals entitled *Self-Service Business Intelligence: TDWI Best Practices Report*[4], The Data Warehousing Institute (TDWI) identified the following top five factors driving businesses toward self-service business intelligence:

- *Constantly changing business needs (65%)*

- *IT's inability to satisfy new requests in a timely manner (57%)*

- *The need to be a more analytics-driven organization (54%)*

- *Slow and untimely access to information (47%)*

- *Business user dissatisfaction with IT-delivered BI capabilities (34%)*

As the business takes greater ownership of its information needs, how does IT's role change? In the same survey report, authors Claudia Imhoff and Colin White suggest that IT's focus shifts toward making it easier for business users "to access the growing number of dispersed data sources that exist in most organizations." Examples include providing friendlier business views of source data; improving on-demand access to data across multiple data sources; enabling data discovery and search functions; supporting access to other types of data, such as unstructured documents; and more.

Given today's information technology challenges, it is time for both IT and the business to move beyond maintaining the status quo and explore together new options to meet increasingly demanding needs for information.

Going Beyond Traditional Data Integration

Over the past fifty years, business use of information systems has expanded from the initial automation of financial accounting functions, such as general ledger and accounts payable, to

---

[4] *Self-Service Business Intelligence: TDWI Best Practices Report*, © TDWI, July, 2011

enterprise-wide business process automation solutions. Examples include enterprise resource planning (ERP), customer relationship management (CRM), human capital management (HCM), supply chain management (SCM) and more.

With core business processes systemized, business intelligence (BI) solutions were a natural follow-on. These solutions enabled business users to leverage the data assets locked inside transaction systems to improve business decision agility and overall business performance.

However, transaction system architectures were optimized for transaction processing, not for the heavy-duty query requirements inherent in BI reporting and analysis solutions. As a result, BI solutions were based on a different architectural paradigm. In this architecture, BI reporting and analysis applications displayed information to business users. Data integration and data management solutions prepared the data behind the scenes.

To support this architecture, new data integration middleware technologies, such as extract, transform and load (ETL), data replication and data propagation, were developed and adopted. And as a complement to this data integration middleware, new data management solutions – e.g., data warehouses, data marts and cubes – emerged to store, manage and deliver the integrated and consolidated data necessary to support BI.

There are many advantages to adopting these now-traditional data integration and data management approaches. The most important is that they enable businesses to successfully meet increasingly complicated information needs. In fact, an entire ecosystem has formed around these approaches to satisfy functionality requirements and reduce risk. Technology vendors provide powerful tools. Organizations such as The Data Warehouse Institute (TDWI) and the Data Management Association (DAMA) provide education and document best practices. Services firms provide external resources to complement internal IT staff.

However, there are two major disadvantages to these traditional approaches. The first disadvantage is the extended time it takes to develop solutions that meet new information requirements and to adapt existing solutions. Because of their inherent architectural complexity, using traditional data integration approaches to support new or changed business needs has typically resulted in long lead times and seemingly endless backlogs. Business dissatisfaction with this slow pace of change, or time to solution, is evidenced in the TDWI survey results shown above. Clearly, this constraint on IT responsiveness is sub-optimal in a dynamic business environment that demands new solutions quickly.

> *Because of their inherent architectural complexity, using traditional data integration approaches to support new or changed business needs has typically resulted in long lead times and seemingly endless backlogs ... Clearly, this constraint on IT responsiveness is sub-optimal in a dynamic business environment that demands new solutions quickly.*

The second disadvantage of using traditional data integration approaches is lack of resource agility. These approaches require design and development in three distinct technologies – BI, data warehousing and ETL. Creating and coordinating metadata, data models, objects and more across these tools is people intensive. Further, replicating data into a data warehouse and/or mart necessitates additional infrastructure and governance resources to effectively manage multiple copies of data. Balancing these resource-intensive efforts against financial constraints often means fewer resources are available to meet new business needs.

Data Virtualization Arises to Improve Agility

To fulfill rapidly-expanding and ever-changing information needs on the business side and at the same time increase time-to-solution and resource agility on the IT side, a new approach to data integration, called *data virtualization*, has evolved with wide adoption over the past ten years.

Data virtualization is a data integration technique that provides complete, high-quality and actionable information through *virtual* integration of data across multiple, disparate internal and external data sources. Data virtualization is implemented using middleware technology that connects to data sources, executes queries to retrieve requested data, combines or federates this data with other data, abstracts and transforms the data to conform to the business information need and then delivers the data to the consuming application.

> *Data virtualization is a data integration technique that provides complete, high-quality and actionable information through virtual integration of data across multiple, disparate internal and external data sources.*

Perhaps the easiest way to understand data virtualization is to contrast it with traditional data integration. Instead of copying and moving existing source data into physical, integrated data stores (e.g., data warehouses and data marts), as is done with traditional data integration approaches, data virtualization creates a virtual or logical data store. In other words, data virtualization leaves source data in place and uses a set of virtual views and data services to access, integrate, represent and deliver the data to business users and applications.

Data virtualization technology includes an integrated development environment (IDE), a data virtualization server environment and a management environment packaged into a complete data virtualization suite. Data virtualization, as a data integration approach and technology, is described more fully in the next chapter, "An Overview of Data Virtualization."

This Is a Book of Data Virtualization Case Studies

This book, the first ever written on the topic of data virtualization, introduces the technology that enables data virtualization and presents ten real-world case studies that demonstrate the significant value and tangible business agility benefits that can be achieved through the implementation of data virtualization solutions.

This first chapter describes the book and introduces the relationship between data virtualization and business agility.

The second chapter is a more thorough exploration of data virtualization technology. Topics include what is data virtualization, why use it, how it works and how enterprises typically adopt it.

The third chapter addresses the many ways that data virtualization improves business agility, with particular focus on the three elements of business agility – business decision agility, time-to-solution agility and resource agility.

 The core of the book is a rich set of in-depth data virtualization case studies that describe how ten enterprises across a wide range of industries and domains have successfully adopted data virtualization to increase their business agility.

The core of the book is a rich set of in-depth data virtualization case studies that describe how ten enterprises across a wide range of industries and domains have successfully adopted data virtualization to increase their business agility. The ten enterprises profiled are customers of Cisco, Inc., a data virtualization software vendor.

These case studies are preceded by a summary chapter entitled "Data Virtualization: The User Perspective" that synthesizes the business agility benefits achieved through data virtualization and provides a compendium of best practices for implementing data virtualization solutions. The summary is followed by the ten detailed case studies (see Figure 1 for an overview).

Each case study includes a complete and rich implementation profile that includes:

• Organization Background – This section provides high-level context about the organization – its industry, products, markets, revenues and more – and a brief profile of the leader of the data virtualization initiative within the organization.

Company	Industry	Domain	Deployment
Comcast	Communications	Directory Services	Ownership change processing
Compassion International	Not-for-profit	Enterprise wide	Ministry Information Library
Fortune 50 Computer Manufacturer	Technology	Procurement	Integrated procurement reporting system
Fortune 50 Financial Services Firm	Financial Services	Wholesale Bank	Support for mergers and acquisitions, new business opportunities
Global 50 Energy Company	Energy	Upstream Operations	Virtual data warehouse to support BI reporting and analytics
Global 100 Financial Services Firm	Financial Services	Investment Bank division	Data Vault
Northern Trust	Financial Services	Corporate and Institutional Services business unit	Investment Operations Outsourcing client reporting platform
NYSE Euronext	Financial Services	Enterprise wide	Virtual data warehouse for post-trade reporting and analysis
Pfizer	Health Care	Worldwide Pharmaceutical Sciences (R&D)	Project portfolio database
Qualcomm	Communications	Enterprise wide	Multiple applications

Figure 1. Overview of Data Virtualization Case Studies

- The Business Problem – This section describes the specific business problem addressed by the data virtualization solution.

- The Data Virtualization Solution – This section describes how data virtualization was successfully deployed to fulfill the business needs described above. This includes information about the data virtualization suite and architecture, the consuming applications and data sources, and the deployment that successfully integrated these sources and consumers. Additional technologies, methodologies and strategies that contributed to the solution are also described.

- The Implementation Process – This section provides insights into the process used to implement the data virtualization solution. We also highlight implementation advice for others based on lessons learned.

- Summary of Benefits and Return on Investment – This section describes the enhanced agility and other business benefits as well as IT benefits realized from the adoption of data virtualization. Where possible, tangible ROI value is included.

- Summary of Critical Success Factors – This section identifies the features of the solution and implementation process that were most important to achieving success.

- Future Directions – Because data virtualization adoption can span multiple projects on the path to enterprise-wide deployment, this section identifies planned next steps as well as other factors to be considered in the near future.

In conclusion, the last chapter of the book transforms earlier chapter insights into a concise set of data virtualization takeaways and guidance to encourage other organizations as they strive to achieve similar business agility success.

Why Read this Book

If your organization is new to data virtualization, the goal of this book is to demonstrate with real-world examples how you can move beyond traditional data integration and use data virtualization to improve your organization's business agility. This requires an understanding of what data virtualization is and a roadmap for its effective implementation.

If your organization is already adopting data virtualization, the goal is this book is to help you successfully accelerate and expand your adoption, compound your business agility gains and achieve other business and IT benefits from data virtualization.

Based on the premise that learning from the experiences of peers is often the most productive way to gain the wisdom and confidence required for success, the case studies are described from the point of view of each organization's data virtualization champion. This provides an insider's perspective that readers can apply to evaluate how data virtualization will work within their organizations.

How to Use this Book

There are many ways to use this book. You can use it as an introduction to the topic of data virtualization. In that case, upon finishing this chapter, drill into a case study or two to get the big ideas before you return to the chapter titled "An Overview of Data Virtualization" to learn the theory behind the practice.

If you already understand data virtualization in general, and are more interested in learning about business value or adoption best practices, start with the chapter titled "Data Virtualization: The User Perspective" where the benefits and best practices are summarized. Then select case studies where the enterprise, industry or business problem is of specific interest to you.

If you are implementing data virtualization now, use this book as a reference for those times when understanding the path taken by others could provide new insights to help you and your organization achieve your business agility objectives sooner.

An Overview of Data Virtualization

Data Virtualization at a Glance

Data virtualization is a data integration approach and technology used by innovative organizations to achieve business agility.

Data virtualization technology is a form of middleware that leverages high-performance software and an advanced computing architecture to integrate and deliver to both internal and external consumers data from multiple, disparate sources in a loosely-coupled, logically-federated manner. By implementing a virtual data integration layer between data consumers and existing data sources, the organization avoids the need for physical data consolidation and replicated data storage (see Figure 1). Thus, data virtualization enables the organization to accelerate delivery of new and revised business solutions while also reducing both initial and ongoing solution costs.

Most front-end business applications, including BI, analytics and transaction systems, can access data through the data virtualization layer. Consumption is on demand from the original data sources, including transaction systems, operational data stores, data warehouses and marts, big data, external data sources and more. High performance query algorithms and other optimization techniques ensure timely, up-to-the-minute data delivery. Logical data models, in the form of tabular or hierarchical schemas, ensure data quality and completeness. Standard APIs and an open architecture simplify the consumer-to-middleware-to-data source connections. Data virtualization middleware suites provide the functionality described above within integrated offerings that support the full software development life cycle, high-performance run-time execution and reliable, 24x7x365 operation.

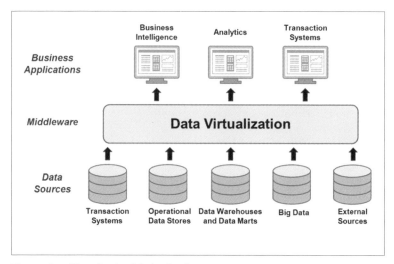

Figure 1. Data Virtualization Solution Landscape
Source: Cisco

How Data Virtualization Technology Works

The primary objects created and used in data virtualization are views and data services. These objects encapsulate the logic necessary to access, federate, transform and abstract source data and deliver the data to consumers. These objects can vary in scope and function depending on the business need, canonical information standards and other usage objectives. Individual objects can call other objects in order to perform additional functions. This is often done using a layered, or hierarchical, approach where objects that perform application delivery functions call objects that perform transformation and conformance functions which, in turn, call objects that perform source data access and validation functions. The ability to reuse common objects in this way provides flexibility, accelerates new development and reduces costs.

 The primary objects created and used in data virtualization are views and data services. These objects encapsulate the logic necessary to access, federate, transform and abstract source data and deliver the data to consumers.

The grouping of objects related to a single domain or subject area, such as trades in financial services or projects in research and development, can be used to create the data virtualization equivalent of a subject-oriented data mart. Multiple domains can then be combined to create the virtual equivalent of a data warehouse. As a result, data virtualization can be adopted in a phased manner, starting with a narrow set of application use cases and expanding over time to a wider, enterprise-scale adoption.

A data virtualization suite consists of three primary middleware components that perform a full range of development, run time and management functions as described below.

INTEGRATED DEVELOPMENT ENVIRONMENT. Data virtualization technology includes an integrated development environment (IDE) that can be used by a range of people, from business analysts to application developers, to define and implement the appropriate view and data service objects. The foundation of these views and services is an underlying logical data model that is, in turn, based on either a tabular or hierarchical schema. Data quality requirements, such as standards conformance, enrichment, augmentation, validation and masking; and security controls (e.g., authentication and authorization) can also be also implemented within these object definitions.

The IDE includes profiling-like introspection and relationship-discovery capabilities designed to simplify a developer's understanding of existing data sources and jump-start the modeling process.

To limit the coding required and save development time, drag-and-drop modeling techniques and a rich set of pre-built, any-to-any transformations automatically generate view or data service objects. Multiple languages (SQL, XQuery, Java, etc.) can extend these capabilities, if required, to address more advanced data virtualization needs. Standard source and consumer APIs, based on ODBC, JDBC, SOAP, REST, etc., simplify source data access and consumer delivery development activities. Integrated data governance, including lineage and where used, metadata asset management and versioning provide needed controls.

 To limit the coding required and save development time, drag-and-drop modeling techniques and a rich set of pre-built, any-to-any transformations automatically generate view or data service objects.

DATA VIRTUALIZATION SERVER ENVIRONMENT. In data virtualization, run-time activities are typically triggered by queries, or requests for data, from a consuming application. The data virtualization server is the component that executes these queries. The query engine within the server, which is specifically designed to process federated queries across multiple sources in a wide-area network, optimizes and executes queries across one or more data sources as defined by the view or data service. Cost- and rule-based optimizers automatically calculate the best query plan for each individual query from a wide variety of supported join techniques. Parallel processing, predicate push-down, scan multiplexing and constraint propagation techniques optimize database and network resources.

The data virtualization server also does the following:

• Transforms query results sets to ensure that the data is complete, high quality and consumable by the user.

• Executes authentication and authorization security functions to protect data from improper use.

- Caches appropriate data sets to enhance both performance and availability.

To complete the query, the server delivers the results directly to the consuming application and logs all activities.

MANAGEMENT ENVIRONMENT. Data virtualization servers are configured for development, testing, staging, production, back-up and failover operations. To manage this topology, meet service-level agreements (SLAs) and ensure reliable 24x7x365 operations, the data virtualization suite also includes a complete set of integrated management tools. These integrated tools support all the activities required to set up the data virtualization middleware and users, including provisioning the software, granting access to sources, integrating with LDAP and other security tools, etc. System management tools manage server sessions and resources. Monitoring tools log activities, monitor memory and CPU usage, as well as display key health indicators in dashboards. Optional clustering tools improve workload sharing and synchronization across servers.

> *The query engine within the server, which is specifically designed to process federated queries across multiple sources in a wide-area network, optimizes and executes queries across one or more data sources as defined by the view or data service.*

DATA VIRTUALIZATION SUITE EXAMPLE. A number of enterprise software vendors provide data virtualization technology. Several of these solutions are delivered as extensions to other technology platforms, such as BI, ETL or an enterprise service bus (ESB). Others, such as the Cisco Data Virtualization Suite from Cisco, are complete, standalone data virtualization suite.

Cisco's key functions are show in Figure 2. IDE functions are on the left, server functions are in the center, and management functions are on the right.

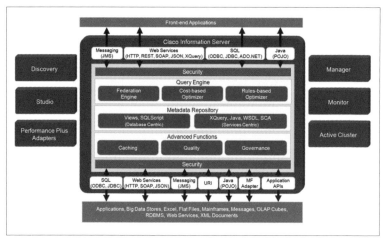

Figure 2. *Cisco Data Virtualization Suite*
Source: Cisco

How Enterprises Adopt Data Virtualization

As with any new technology, organizations begin the data virtualization adoption process with a business justification and a technical evaluation, followed by a phased implementation.

BUSINESS JUSTIFICATION. As described above, the business value of data virtualization comes from its ability to help organizations deliver complete, high-quality and actionable information more quickly and with fewer resources than traditional data integration approaches. This faster time-to-solution advantage translates into faster realization of the business benefits – e.g., increased revenue, improved customer service and retention, enhanced competitive responsiveness and better regulatory compliance – that are the business drivers behind new information requests.

An example is the best way to understand why benefits are realized faster with data virtualization. Let's assume the business wants to implement an enhanced customer self-service portal in order to improve customer satisfaction. This initiative is financially justified based on the expectation that it will generate one million dollars in additional revenue per month. If the use of data virtualization would result in a three-month shorter time to solution than an alternative data integration approach, then an additional three million dollars would be generated.

Additional benefits would include initial IT development resource savings from the faster development time and long-run infrastructure cost avoidance due to reduced data duplication. This combination of hard and soft savings could then be used to justify the purchase of data virtualization technology.

TECHNICAL EVALUATION. Once the business case is clear, a thorough technical evaluation of alternative data virtualization offerings is often the next step in the process of adoption. IT organizations typically have standard methodologies for identifying a set of viable technology vendors, refining it to a short list for detailed evaluation and then performing deep-dive proofs of concept and "bake-offs" that result in the selection and purchase of an appropriate data virtualization technology offering.

PHASED ADOPTION. As described in a number of the case studies, once a data virtualization technology has been purchased, most organizations initially deploy data virtualization on a subset of possible data integration projects, functional groups and/or information domains. An example is BI for research and development. This allows the organization to concentrate expertise and thus be more effective in its initial use of data virtualization technology. From this foundation, within a year or two, deployments typically expand to more use cases across more domains and groups, eventually resulting in an enterprise-scale deployment.

Five Popular Data Virtualization Usage Patterns

Data virtualization is a versatile data integration solution that can be deployed to solve a wide range of data integration challenges. Based on nearly ten years of successful implementations, several common usage patterns have emerged to help guide your enterprise's data virtualization adoption strategy.

- BI data federation

- Data warehouse extension

- Enterprise data virtualization layer

- Big data integration

- Cloud data integration

BI DATA FEDERATION. Historically, BI data federation has been the most popular data virtualization starting point. Data virtualization is an excellent way to expand BI reporting to sources beyond those captured in existing cubes or data marts. Popular BI vendors have promoted this approach and even embedded data virtualization offerings in their BI solutions to simplify this method of adoption. The benefits of this approach include both more complete and actionable information and faster time-to-solution.

DATA WAREHOUSE EXTENSION. Extending the data warehouse has gained popularity in recent years as another starting point for adopting data virtualization. In the struggle to maintain data warehouses and keep pace with accelerating business and technology changes, IT organizations have adopted data virtualization as an effective way to augment warehouse data. Examples include augmenting internal warehouse data with external data, combining last night's warehouse data with today's data and complementing summarized warehouse data with detailed drill-down data.

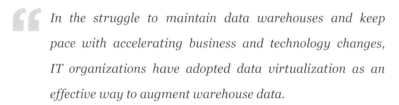

In the struggle to maintain data warehouses and keep pace with accelerating business and technology changes, IT organizations have adopted data virtualization as an effective way to augment warehouse data.

Data warehouse prototyping, federating multiple warehouses after a merger and support for data warehouse migration are additional sub-patterns within the data warehouse extension approach. Beyond information and agility value, enterprises also increase their return on prior data warehouse investments.

ENTERPRISE DATA VIRTUALIZATION LAYER. Once comfortable with the initial, project-oriented data virtualization deployments, the next step is often deploying a more enterprise-wide data virtualization layer. A data virtualization layer combines SOA principles, such as decoupling, reuse, and agility, with key information governance principles, such as abstraction, shared semantic models and data standards, to enable IT to build and deploy a layered, enterprise-wide data architecture in a simpler, faster, more consistent and scalable manner. From the business point of view, this layer serves as a unified, enterprise-wide view of business information that improves users' ability to understand and leverage enterprise data.

BIG DATA INTEGRATION. Big data integration has recently emerged as a popular data virtualization usage pattern driven by enterprises' rapid adoption of analytics. Most of these analytics run on new "Big Data" data stores such as Hadoop and analytic data warehouse appliances such as IBM Netezza. Once these new stores are in place, organizations soon realize they can gain additional insight if they integrate these new big data silos with their existing enterprise data. This is where data virtualization comes in, providing a rapid integration approach that doesn't require additional replication of already "big" data sources.

CLOUD DATA INTEGRATION. Cloud data integration has also recently emerged as a popular data virtualization usage pattern as enterprises take advantage of new Software-as-a-Service (SaaS), Platform-as-a-Service (PaaS), and Infrastructure-as-a-Service (IaaS) offerings. However, each new cloud source and consumer must be integrated with the existing IT environment – a problem that data virtualization is ideally architected to solve. This allows enterprises to maintain a complete view of their internal and external information while taking advantage of attractive cloud economics.

Data Virtualization and Competency Centers

In many large enterprises, a BI or Integration Competency Center (ICC) is often the nexus for enterprise-wide scale-out of technology such as data virtualization. ICCs combine and evolve the people, processes and technologies required to maximize both business and IT benefits. Including data virtualization within an existing ICC enables the organization to achieve the following:

- Better leverage the existing ICC infrastructure, organizational alignment and lessons learned to accelerate data virtualization's IT agility and cost saving benefits.

- Help guide a unified, complete data virtualization strategy and architecture to support rapid execution, concentrate subject-matter expertise for faster resolution of data virtualization design and development issues, support object reuse and accelerate IT agility.

- Help reduce initial and ongoing costs by optimizing internal and external staffing across multiple data virtualization projects and activities. This improves use of the data virtualization infrastructure, facilitates the development and consistent use of best practice methods and enables the organization to more effectively anticipate and proactively address problems to avoid remediation costs.

Bringing data virtualization into an existing ICC need not be difficult. Yet every situation will be different. Successful organizations adopt a flexible implemention approach adapted to their unique situations and needs. When developing a plan, the organization should consider multiple factors including data virtualization maturity, degree of resource centralization versus decentralization, staff skills, alignment with other competency centers, etc. Starting small by limiting the scope of functions or projects can reduce risk. It is also important to communicate the accomplishments of the data virtualization group within the ICC, as this helps data virtualization gain momentum and accelerate success.

How Data Virtualization Delivers Business Agility

Business Agility Challenges

Today's business environment is dynamic, complex and challenging due to a myriad of internal strategic, operational and financial objectives plus external competitive and regulatory compliance hurdles. To successfully run the business today while at the same time adapting it to meet new challenges tomorrow requires access to complete, high-quality and actionable information.

As an example, take the recent acceleration in the adoption of predictive analytics. In their best-selling book *Competing on Analytics: The New Science of Winning*[1], authors Thomas H. Davenport and Jeanne G. Harris *"found a striking relationship between the use of analytics and business performance ... High performers (those who outperformed their industry in terms of profit, shareholder return and revenue growth) were 50 percent more likely to use analytics strategically ... and five times as likely as low performers."*

Predictive analytics opportunities, which include pricing optimization, sales and inventory forecasting, customer churn

---

[1] *Competing on Analytics: The New Science of Winning*, Thomas H. Davenport and Jeanne G. Harris, © Harvard Business School Publishing Corporation, 2007

prevention, marketing campaign optimization, fraud detection and supply chain management, were relatively nascent just few years ago. They are now both abundant and critical for business success.

However, while the value of analytics is unquestioned, a number data integration issues often slow the adoption of analytics and thus delay business benefits.

• Data complexity – Effective analytics applications often must leverage data from multiple, diverse internal and external sources, including relational, semi-structured XML, dimensional MDX and new "Big Data" data types supported by platforms such as Hadoop and data warehouse appliances. Frequently, however, the characteristics of the data across these sources – e.g., completeness, structure, syntax, availability and quality – do not meet the requirements of these new analytics applications.

• Query performance – Large volumes of data must be analyzed and access to up-to-the-minute information is often important when making critical business decisions. These factors make query performance a critical success factor for data integration efforts.

• IT responsiveness – Dynamic businesses require new and ever-changing analyses. This means new data sources must be brought on board quickly and existing sources must be modified to support each new analytic requirement.

This predictive analytics example demonstrates the important role of data integration in supplying business decision makers with the complete, high-quality and actionable information they need.

Business Agility Requires Data Integration Flexibility

The need for a flexible approach to data integration is accelerating as information usage expands to support additional business users with varied roles and responsibilities, and an increasingly diverse and extended range of analysis and reporting applications that address new and different business problems.

On the IT side, the number of data sources is also growing. Each new source adds volume, variety, velocity and complexity to a data integration landscape that already must incorporate generations of legacy infrastructure as well as new architectures based on cloud and mobile computing paradigms.

Accenture expects continued acceleration in these areas in its report *Accenture Technology Vision 2011*[2]. In particular, the report describes three key business and IT trends driving the need for flexible data integration:

*"**Things will be distributed** – The obvious and immediate realization is that data today is spread far and wide. Data is also dispersed across many more locations, and under the control of far more owners. At the same time, services will be distributed more widely. Analytics will follow data and services, and will become distributed too.*

***Things will be decoupled** – Technology today enables decoupling of entities and layers once deemed inseparable. Data representation is being decoupled from data access. Software layers can be addressed separately. Application interfaces no longer need to be tied to physical interfaces. Decoupling on such a scale promises unprecedented agility and flexibility. But it also calls for a very different mindset – and skills set – and for wise governance disciplines.*

***Things will be analyzed** – Since everything from keystrokes to consumer behavior can now be tracked and studied, analytics will become the super-tool with which to drive more agile and effective decision-making. Business processes will have to keep pace if those super-tools are to be effective."*

These drivers challenge the validity of data integration architectures in which a single, enterprise-wide data warehouse attempts to serve as the repository for all enterprise data. Instead, analyst firms now recommend, and enterprise architects embrace, the need for a broad portfolio of data integration approaches that includes data

---

[2] *Accenture Technology Vision 2011, © Accenture, 2011*

virtualization along with traditional data integration techniques. Traditional approaches encompass data consolidation in the form of data warehouses and marts enabled by ETL as well as messaging- and replication-based approaches that move data from one location to another.

This broader portfolio offers IT the flexibility to mix and match data integration techniques and technologies according to the unique requirements of the organization's information consumers, the key characteristics of its data sources and other important factors, such as optimal time to solution, development and maintenance cost and data latency.

Business Agility Needs Drive Data Virtualization Adoption

Enterprise adoption of data virtualization has accelerated along with a growing need for greater business agility. This relationship can be observed across hundreds of organizations and is clearly evident in the ten case studies described in this book.

Data virtualization is ideally suited to provide the level of business decision agility, time-to-solution agility and resource agility so critical to successfully achieving overall business agility. How data virtualization fulfills these business agility needs and the unique advantages data virtualization offers over other data integration approaches are described below.

Business Decision Agility

Making effective business decisions requires knowledge and insight that can only be developed from access to and analysis of complete, high-quality, actionable information. Data virtualization enables the organization to deliver this information in several ways.

COMPLETE INFORMATION. Understanding the complete picture is the first step in any decision process. Large enterprises today have thousands of data sources that span multiple transaction systems of record, complementary applications, consolidated data stores, external data sources and more. Each source is a silo with its own unique metadata and data model, data access toolset, and underlying architecture. The challenge is integrate data across these traditional silos in order to provide the business user with a single, complete and high-level view of whatever information is needed for analysis and decision making. Taking this concept to the enterprise level, data virtualization has the ability to provide an organization with a unified view of information across the entire business.

One traditional solution is to consolidate all of the data in a unified, enterprise data warehouse. However, this does not always prove feasible in practice for a number of reasons, including the ongoing proliferation of new data sources and types that must be incorporated into the warehouse.

Data virtualization has the ability to provide an organization with a unified view of information across the entire business.

As an alternative, data virtualization offers virtual data federation functions that enable an organization to integrate its extensive range of internal and external data sources without moving any data. Accessing the data in a new source, for example, simply requires establishing a single connection – from the data virtualization layer to the source – and the creation of virtual views and data services to access, transform, abstract and represent the source data in an appropriate format for consuming applications.

HIGH-QUALITY INFORMATION. Ensuring that the information guiding business decisions is high quality and fit-for-purpose is a second major business decision agility requirement. It is rare that as-is, raw data in original source systems is an exact match for the consuming application and business user. At a minimum, some degree of format and syntax transformation is required to bridge the gap between source and consumer data models and technologies. In many cases, additional validation and standards-conformance processing is also needed to improve the integrity and consistency of the data delivered to consumers. Data virtualization supports multiple techniques to ensure delivery of high-quality information. For example, when providing up-to-the-minute, institution-wide views of equity, option, futures, derivative and debt positions for risk managers in the financial services industry, data virtualization transforms the data as structured in the various trading sources into a consistent form and format for consumption by the risk management applications.

ACTIONABLE INFORMATION. Finally, the information decision makers require must be actionable. Time-to-action is an additive function that combines time from event-to-insight, insight-to-decision, decision-to-implementation and implementation-to-results. Yesterday's data, summarized in the warehouse, is not sufficient when having the most current information is the first step in a time-to-action path. For example, understanding the current location and availability of maintenance staff and repair gear is a critical first step in understanding how to respond to an equipment failure in process industries. Data virtualization provides high-performance query engines and flexible caching to query and deliver source data in near-real time whenever it is requested. This ensures decisions are made based on most up-to-date information available when appropriate.

Time-To-Solution Agility

When responding to new information needs, rapid time to solution is critically important and often results in significant bottom-line benefits. Consider this example. If the business wants to enter a new market, it must first financially justify the investment, including any new IT requirements. Thus, only the highest ROI projects are approved and funded. Once the effort is approved, accelerating delivery of the IT solution also accelerates realization of the business benefits and ROI. Therefore, if incremental revenues from the new market are $2 million per month, then the business will gain an additional $2 million for every month IT can save in time needed to deliver the solution.

STREAMLINED APPROACH. Data virtualization is significantly more agile and responsive than traditional data consolidation and ETL-based integration approaches because it uses a highly streamlined architecture and development process to build and deploy data integration solutions. This approach greatly reduces complexity and reduces or eliminates the need for data replication and data movement. As all of the case studies demonstrate, this elegance of design and architecture makes it far easier and faster to develop and deploy data integration solutions using a data virtualization suite. The ultimate result is faster realization of business benefits.

> *Data virtualization is significantly more agile and responsive ... because it uses a highly streamlined architecture and development process to build and deploy data integration solutions.*

To better understand the difference, let's contrast these methods. In both in the traditional data warehouse/ETL approach and data virtualization, understanding the information requirements and reporting schema is the common first step.

Using the traditional approach, IT then models and implements the data warehouse schema. ETL development follows to create the links between the sources and the warehouse. Finally the ETL scripts are run to populate the warehouse. The metadata, data models/schemas and development tools used within each activity are unique to each activity. This diverse environment of different metadata, data models/schemas and development tools is not only complex but also results in the need to coordinate and synchronize efforts and objects across them. Experienced BI and data integration users will readily acknowledge the long development times that result from this complexity, including Forrester Research in its 2011 report *Data Virtualization Reaches Critical Mass*[3]:

---

[3] *Data Virtualization Reaches Critical Mass, © Forrester Research, June, 2011*

"Extract, transform, and load (ETL) approaches require one or more copies of data staged along the physical integration process flow. Creating, storing, and manipulating these copies can be complex and error prone."

Data virtualization uses a more streamlined architecture that simplifies development. Once the information requirements and reporting schema are understood, the next step is to develop the objects (views and data services) used to both model and query the required data. These virtual equivalents of the warehouse schema and ETL routines and scripts are created within a single view or data service object using a unified data virtualization development environment. This approach leverages the same metadata, data models/schemas and tools. Not only is it easier to build the data integration layer using data virtualization, but there are also fewer "moving parts," which reduces the need for coordination and synchronization activities.

With data virtualization, there is no need to physically migrate data from the sources to a warehouse. The only data that is moved is the data delivered directly from the source to the consumer on demand. These result sets persist in the data virtualization server's memory for only a short interval. Avoiding data warehouse loads, reloads and updates further simplifies and streamlines solution deployment and thereby improves time-to-solution agility.

ITERATIVE DEVELOPMENT PROCESS. Another way data virtualization improves time-to-solution agility is through support for a fast, iterative development approach. Here, business users and IT collaborate to quickly define the initial solution requirements followed by an iterative "develop, get feedback and refine" process until the solution meets the user need.

Most users prefer this type of development process. Because building views of existing data is simple and fast, IT can provide business users with prospective versions of new data sets in just a few hours. The user doesn't have to wait months for results while IT develops detailed solution requirements. Then business users can react to these data sets and refine their requirements based on the tangible insights. IT can then change the views and show the refined data sets to the business users. This iterative development approach enables the business and IT to home in on and deliver the needed information much faster than traditional integration methods.

Even in cases where a data warehouse solution is mandated by specific analytic needs, data virtualization can be used to support rapid prototyping of the solution. The initial solution is built using data virtualization's iterative development approach, with migration to the data warehouse approach once the business is fully satisfied with the information delivered.

In contrast, developing a new information solution using a traditional data integration architecture is inherently more complex. Typically, business users must fully and accurately specify their information requirements prior to any development, with little change tolerated. Not only does the development process take longer, but there is a real risk that the resulting solution will not be what the users actually need and want. Data virtualization offers significant value, and the opportunity to reduce risk and cost, by enabling IT to quickly deliver iterative results that enable users to truly understand what their real information needs are and get a solution that meets those needs.

EASE OF CHANGE. The third way data virtualization improves time-to-solution agility is ease of change. Information needs evolve. So do the associated source systems and consuming applications. Data virtualization allows a more loosely-coupled architecture between sources, consumers and the data virtualization objects and middleware that integrate them. This level of independence makes it significantly easier to extend and adapt existing data virtualization solutions as business requirements or associated source and consumer system implementations change. In fact, changing an existing view, adding a new source or migrating from one source to another is often completed in hours or days, versus weeks or months in the traditional approach.

Resource Agility

While the benefits derived from greater business agility are significant, costs are also an important factor to consider. This is especially true in today's extremely competitive business environment and difficult economic times. As a result, resource agility is the third key element in an enterprise's business agility strategy.

Fortunately, data virtualization directly enables greater resource agility through superior developer productivity, lower infrastructure costs and better optimization of data integration solutions. These factors combine to provide significant cost savings that can be applied flexibly to fund additional data integration activities and/or other business and IT projects.

SUPERIOR DEVELOPER PRODUCTIVITY. At 41% of the typical enterprise IT budget, personnel staffing expenses, including salaries, benefits and occupancy, represent the largest category of IT spending[4]. This spending is double that of both software and outsourcing, and two-and-a-half times that of hardware. Not only are these staffing costs high in absolute terms. With data integration efforts often representing half the work in a typical IT development project, data integration developer productivity is critically important on a relative basis as well.

As described above, data virtualization uses a streamlined architecture and development approach. Not only does this improve time-to-solution agility, it also improves developer productivity in several ways.

• First, as discussed above, data virtualization allows rapid, iterative development of views and data services. The development and deployment time savings associated with this development approach directly translate into lower staffing costs.

• Second, the typically SQL-based views used in data virtualization are a well-understood IT paradigm. And the IDEs for building these views share common terminology and techniques with the

---

[4] *IT Metrics: IT Spending and Staffing Report, © Gartner, Inc., January, 2011*

IDEs for most popular relational databases. The same can be said for data services and popular SOA IDEs. These factors make data virtualization easy for developers to learn and reduce training costs typically required when adopting new tools.

- Third, graphically-oriented IDEs simplify data virtualization solution development with significant built-in code generation and automatic query optimization. This enables less senior and lower cost development staff to build data integration solutions.

- Fourth, the views and services built for one application can easily be reused across other applications. This further increases productivity and reduces staffing resource costs.

LOWER INFRASTRUCTURE COSTS. Large enterprises have hundreds, if not thousands, of data sources. While these data assets can be leveraged to provide business decision agility, these returns come at a cost. Each source needs to be efficiently operated and managed and the data effectively governed. These ongoing infrastructure costs typically dwarf initial hardware and software implementation costs.

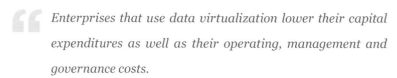

Enterprises that use data virtualization lower their capital expenditures as well as their operating, management and governance costs.

Traditional data integration approaches, where data is consolidated in data warehouses or marts, add to the overall number of data sources. This necessitates not only greater up-front capital expenditures, but also increased spending for ongoing operations and management. In addition, every new copy of the data introduces an opportunity for inconsistency and lower data quality. Protecting against these inevitable issues is a non-value-added activity that further diverts critical resources. Finally, more sources equal more complexity. This means large, ongoing investments in coordination and synchronization activities.

These demands that consume valuable resources that can be significantly reduced through the use of data virtualization. Because data virtualization requires fewer physical data repositories than traditional data integration approaches, enterprises that use data virtualization lower their capital expenditures as well as their operating, management and governance costs. In fact, many data virtualization users find these infrastructure savings alone can justify their entire investment in data virtualization technology.

OPTIMIZED DATA INTEGRATION. As a component of a broad data integration portfolio, data virtualization joins traditional data integration approaches such as data consolidation in the form of data warehouses and marts enabled by ETL as well as messaging- and replication-based approaches that move data from one location to another. Each of these approaches has strengths and limitations when addressing various business information needs, data source and consumer technologies, time-to-solution and resource agility requirements.

For example, a data warehouse approach to integration is often deployed when analyzing historical time-series data across multiple dimensions. Data virtualization is typically adopted to support one or more of the five usage patterns described in the prior chapter:

• BI data federation

• Data warehouse extension

• Enterprise data virtualization layer

• Big data integration

• Cloud data integration

Given the many information needs, integration challenges, and business agility objectives organizations have to juggle, each data integration approach added to the portfolio improves the organization's data integration flexibility and thus optimizes the ability to deliver effective data integration solutions.

With data virtualization in the integration portfolio, the organization can optimally mix and match physical and virtual integration methods based on the distinct requirements of a specific application's information needs, source data characteristics and other critical factors such as time to solution, data latency and total cost of ownership.

In addition, data virtualization provides the opportunity to refactor and optimize data models that are distributed across multiple applications and consolidated stores. For example, many enterprises use their BI tool's semantic layer and/or data warehouse schema to manage data definitions and models. Data virtualization provides the option to centralize this key functionality in the data virtualization layer. This can be especially useful in cases where the enterprise has several BI tools and/or multiple warehouses and marts, each with their own schemas and governance.

Data Virtualization: The User Perspective

Case Study Overview

For this book, we interviewed representatives of ten organizations that have implemented data virtualization solutions to improve their overall business agility. Our goal was to understand the challenges of data virtualization, how solutions were developed, the benefits achieved from the user's perspective and the best practices advice these experienced implementers could offer to other organizations with similar business problems. The following chapters present in-depth data virtualization case studies that cover, for each organization, the specific business problem, the data virtualization solution, implementation process and advice, benefits achieved and ROI, critical success factors and future directions.

In this chapter, we provide a summary of the significant benefits these organizations have achieved from their data virtualization efforts and relate the benefits to those we described in earlier chapters. In addition, we have also developed a set of best practices for implementing data virtualization. These are derived from the implementation advice and description of critical success factors graciously offered by these data virtualization experts.

We encourage you to also read the detailed case studies. They offer a wealth of information about what it is like to design, implement and manage a successful data virtualization solution.

The Users

The ten case study organizations represent a diverse set of industries, including communications, energy, financial services, health care, not for profit, and technology (see Figure 1). The scope of their solutions span specific data virtualization projects, projects with a division/business unit focus and development of an enterprise-wide data virtualization layer. Those organizations that have not already planned an enterprise-level implementation are clear on the benefits of extending their use of data virtualization in new areas going forward.

Company	Industry	Domain	Deployment
Comcast	Communications	Directory Services	Ownership change processing
Compassion International	Not-for-profit	Enterprise wide	Ministry Information Library
Fortune 50 Computer Manufacturer	Technology	Procurement	Integrated procurement reporting system
Fortune 50 Financial Services Firm	Financial Services	Wholesale Bank	Support for mergers and acquisitions, new business opportunities
Global 50 Energy Company	Energy	Upstream Operations	Virtual data warehouse to support BI reporting and analytics
Global 100 Financial Services Firm	Financial Services	Investment Bank division	Data Vault
Northern Trust	Financial Services	Corporate and Institutional Services business unit	Investment Operations Outsourcing client reporting platform
NYSE Euronext	Financial Services	Enterprise wide	Virtual data warehouse for post-trade reporting and analysis
Pfizer	Health Care	Worldwide Pharmaceutical Sciences (R&D)	Project portfolio database
Qualcomm	Communications	Enterprise wide	Multiple applications

Figure 1. Overview of Data Virtualization Case Studies

Benefits of Data Virtualization

All ten organizations clearly emphasize enhanced business agility as the primary overall goal and benefit of their data virtualization efforts. The actual benefits achieved are impressive and should encourage other organizations to consider and actively pursue data virtualization opportunities.

Specific benefits are summarized below within the appropriate element of business agility – business decision agility, time-to-solution agility and resource agility. Every organization described benefits in at least two of these components, and all have achieved resource agility benefits. Three companies have realized benefits in all three areas.

BUSINESS DECISION AGILITY. Multiple users cited the ability to deliver complete, high-quality and actionable information to data consumers to support improved business decisions as a major plus for data virtualization. Here are some examples.

Better decisions, reduced risk, easier identification of business opportunities – The Global 50 Energy Company stated that faster delivery of integrated, high-quality information supports better business decisions that in turn reduce risk. In particular, integration of key data makes it easier for the business to identify and act on opportunities to increase revenue and reduce costs. Examples include making decisions about where to drill new wells and comprehensive reporting to better control global supplier costs.

The Global 100 Financial Services Firm cited a similar benefit, that easy, real-time access to a high-quality, comprehensive view of the business enables users to develop market insights and make more effective investment and risk management decisions.

The Fortune 50 Financial Services Firm highlighted the potential to realize additional revenue streams through the creation of data services within the data virtualization layer. These services enable the business to monetize its data and applications to tap into previously unavailable revenue streams (e.g., to expand its distribution channels for a product or service).

High-quality, integrated data in context – For Compassion International, data virtualization has improved both the quality and integrity of data to produce consistent results on reports. This is critical to the organization's ability to continue good stewardship of its mission.

By using a data virtualization layer to integrate across silos of data, Pfizer now can provide integrated data in context for R&D staff, adding significant value to the information.

More efficient analysis and decision making – Three companies mentioned the ability of data virtualization to improve their business decision agility by delivering comprehensive information in an actionable format. This dramatically reduces the time decision makers have to spend on "non-value-add" activities, such as manually pulling together the information needed for analysis and reporting. With data virtualization, the Global 50 Energy Company can "do things faster and smarter." This results not only in a competitive edge, but also improves efficiency and facilitates business resource reallocation. An example is the elimination of time spent compiling reports manually. This frees up time for users to reallocate time and effort to more valuable business activities.

Pfizer has experienced this benefit as well, citing the elimination of significant manual effort throughout the PharmSci group with delivery of comprehensive, integrated data through their data virtualization suite. One example is a reduction in the number of meetings needed to reconcile data. The Global 100 Financial Services Firm described a similar scenario, where business users can now focus their time and effort on analysis activities rather than on finding and reconciling data.

TIME-TO-SOLUTION AGILITY. An important benefit of data virtualization is the ability to deliver new and revised solutions to users much faster than is possible with traditional data integration approaches. Quick development and delivery of solutions is facilitated by many features of a data virtualization environment, including a streamlined approach to data integration, support for an iterative development process and ease of change within the data virtualization environment. A majority of the users we interviewed stated that time-to-solution agility was a key factor in their decision to implement data virtualization.

A streamlined approach to data integration – The Fortune 50 Financial Services Firm implemented data virtualization to meet critical time constraints for integrating data during a merger. The solution enabled the company to immediately integrate data in areas where it was important to present a seamless transition to customers of the two merging organizations. Success was critical to retain customers in a competitive environment, and using a traditional data integration approach would have taken too long.

An important benefit of data virtualization is the ability to deliver new and revised solutions to users much faster than is possible with traditional data integration approaches.

Northern Trust found that data virtualization enables it to dramatically reduce the time to implement a new outsourcing customer. The company has experienced a 50% reduction in time to market and a 200% ROI. In addition, the faster time to market increases customer satisfaction and revenue.

The Fortune 50 Computer Manufacturer also mentioned faster time to solution as a plus for data virtualization. The company was able to quickly deliver a more flexible procurement reporting system and a globalized view of procurement data that leveraged multiple existing procurement databases. The company saved over $1 million in development and infrastructure costs by delivering the solution faster with data virtualization. It also saved millions of dollars through faster inventory turns and improved customer satisfaction.

The Global 50 Energy Company was able to reduce its time to solution through several features of data virtualization. One is the simpler, more agile architecture offered by a data virtualization approach.

Another is that data virtualization provides a more readily accessible integration environment that supports reporting and analytical tool flexibility. The Energy Company also found that it could ensure high performance by effectively and easily incorporating massively parallel processing technology within the data virtualization environment.

Iterative development process – Several companies mentioned the benefits of using an iterative development process to achieve faster turnaround on new requests for data. Compassion International likes the fact that this contributes to a leaner development process, enables IT to deal with the biggest risks earlier in the process and essentially "fail faster," reducing time to solution by more than 50%.

> *With an iterative process, IT can get feedback to customers faster and model applications in a few days instead of spending a lot of time developing detailed specs and requirements.*

This concept of "fail faster" is a critical element of the iterative development process and was also mentioned by Qualcomm and Pfizer. With an iterative process, IT can get feedback to customers faster and model applications in a few days instead of spending a lot of time developing detailed specs and requirements. Here, Pfizer stressed the importance of being able to provide results in a timeframe relevant to the user. Features of the data virtualization environment which support an iterative development process to deliver solutions faster include reuse of shared services, easy creation of virtual views, the building block approach to developing a hierarchy of virtual views and the ability to rapidly and easily prototype solutions.

Ease of change – Four organizations – Northern Trust, NYSE Euronext, Pfizer and the Global 100 Financial Services Firm – highlighted benefits from the fact that the data virtualization layer makes applications and sources independent of each other by abstracting the data and the physical implementation of the architecture. Thus, IT can make changes to back-end data sources (e.g., swapping out database platforms or changing database structures) without affecting front-end applications. Data virtualization ensures that change management is easy for IT and transparent to users. Another example is migrating from one database to another while maintaining parallel operation of both the legacy and new data sources. The user application doesn't need to know which data source it is using.

Pfizer also emphasized the ability to embed all business rules in the data virtualization layer with the ability to manage and monitor everything in one centralized location.

RESOURCE AGILITY. Based on the case studies, data virtualization delivers huge benefits in the area of increased resource agility through superior developer productivity, lower infrastructure costs and optimized data integration. All ten organizations have achieved significant cost savings and/or opportunities for resource reallocation through the implementation of data virtualization solutions.

Superior developer productivity – Several organizations highlighted features of data virtualization that enhance developer productivity by simplifying application development and maintenance and thus reducing overall costs. Northern Trust, for example, stated that data virtualization provides a single data access point for all consuming applications and the ability to define data services that are reusable by multiple applications. Both features significantly reduce application complexity and development and maintenance time and cost for the company.

The Global 100 Financial Services Firm echoed the importance of a standardized access point for data, focusing on the reduced cost and time to develop new applications and reporting systems. This company's data virtualization solution also achieves more cost-effective and efficient use of application development resources by allowing only one copy of source data to be made (into the central Data Vault repository). Thus, the company avoids the need to devote application resources to maintaining multiple extracts of data and the need to then reconcile multiple versions of the same data. This enables application teams to redirect resources to more valuable development activities.

The Global 50 Energy Company also mentioned the benefits of code reuse on both the front and back end, stating that data virtualization has enabled the organization to reduce its overall development costs by 40%. The company also finds that a data virtualization approach requires far fewer resources to maintain and support the data integration environment than an ETL approach.

Comcast's data virtualization solution takes advantage of the ability to push the transaction management function down from the application layer to the data virtualization layer, again simplifying and reducing the cost of application development and maintenance and eliminating the need for complex reconciliation of transaction data.

Compassion International mentioned two additional features that enhance the flexibility of the data virtualization suite. One is that all data sources appear as a single view of the data, which makes it easier for IT to deliver data in multiple ways to different consumers. Another is that, with data virtualization, the organization standardizes business logic and terms through the virtual views, not through ETL processes. This simplifies maintenance of business logic and makes development of new solutions easier. With its data virtualization solution, Compassion has reduced the total cost of developing data services and reporting and analysis applications by 25 to 30% per project. This frees up developer time to complete more projects and focus on making more useful information available to the organization.

Three organizations also noted the importance of SQL support within the data virtualization development environment. This enables the organization to leverage existing skill sets. For Qualcomm, the fact that data virtualization does not introduce a new development language makes the data virtualization suite easy to use with very little training for existing report developers and database programmers who are familiar with SQL. The Fortune 50 Computer Manufacturer stated that support for SQL enabled its IT team to maintain high productivity and minimize staff disruption. Comcast added that because its data virtualization deployment supports access to LDAP data via SQL, developers trained in SQL are now able to more easily access LDAP data

Lower infrastructure costs – Several organizations also indicated ways in which a data virtualization approach to data integration reduces overall data integration infrastructure costs. The Global 50 Energy Company said that its data virtualization environment had an overall lower total cost of ownership (TCO), including minimal requirements for support staff as described above.

The Fortune 50 Computer Manufacturer stated that its data virtualization suite avoids the significant infrastructure costs associated with physical integration of data, including redundant data and storage. The company has also eliminated potential security risks and reduced the cost of ongoing maintenance by migrating a legacy reporting system over to data virtualization.

With data virtualization, Qualcomm no longer has to store, manage and synchronize the same data in multiple locations. Because ownership is well-defined, the company can establish stronger ownership of data as part of its governance effort.

The Fortune 50 Financial Services Firm considers data virtualization a cost-effective infrastructure approach for data delivery because it consolidates the software licensing and physical infrastructure associated with two distinct technology categories. One is building data integration solutions. Here, data virtualization includes caching and minimizes the need for a separate ETL infrastructure. The second is building a services layer and APIs around data. In this case, data virtualization doesn't require separate application and web servers, which reduces cost and increases cost-effectiveness.

NYSE Euronext emphasized the fact that data virtualization reduces the footprint of software deployed through the reuse of common data services and infrastructure. Data virtualization makes common logic reusable by multiple consumers with no need to propagate logic changes to applications. Thus, the organization can push common functions that have nothing to do with the business into the shared middle tier. Examples are how data is accessed and managed, performance algorithms, high availability, backup and recovery and caching support. This enables the organization to make the system faster and more resilient to better meet SLAs while avoiding duplicate development costs. NYSE Euronext saved over $4.5 million annually by migrating one legacy, outsourced system to its data virtualization environment.

> *Data virtualization makes common logic reusable by multiple consumers with no need to propagate logic changes to applications.*

Optimized data integration – The flexibility of data virtualization to integrate data while at the same time accommodating existing real world, legacy systems is key to the overall success of the approach. Here are several examples. The Fortune 50 Computer Manufacturer describes its global procurement reporting system as an "optimal" data integration solution because, with data virtualization, the company was able to implement the desired level of globalization at both the BI and data levels while maintaining the regional application instances, which were important for isolation, performance and legacy reasons.

NYSE Euronext and the Global 50 Energy Company have both been able to combine data virtualization technology with massively parallel processing technology to optimize the performance of their data virtualization architectures.

By adding data virtualization to its data integration portfolio, Qualcomm has reduced the need to rely solely on traditional integration technologies. When data virtualization is the appropriate solution, the company avoids investing the extra time needed to support more complex ETL integration projects, synchronize and reconcile multiple copies of data, etc. The company estimates it has saved over $2.2 million in development costs on its five initial data virtualization projects.

Best Practices for Implementing Data Virtualization

All of these user organizations have been involved in evaluating and implementing data virtualization solutions for at least two years, some for as long as four to six years. The lessons learned along the way are valuable to pass on to help other organizations avoid common pitfalls and realize the benefits of data virtualization as quickly as possible. We asked the representatives in our case studies to describe the aspects of their solutions and implementation processes that were critical to success and to share any implementation advice they would give others in a similar situation. Here is a summary of what they said.

Centralize responsibility for implementing data virtualization – The Global 50 Energy Company stressed the need to centralize the initial design, development and deployment responsibility for data virtualization into a focused data virtualization team. The key benefit here is the ability to advance the effort quickly and to take on the bigger concepts, such as defining common canonicals and implementing an intelligent storage component to speed development, reduce time to solution and deliver a more powerful and complete data virtualization environment. This company also separated responsibility for architecture and development within its data virtualization team.

Northern Trust echoed this advice to centralize support for data virtualization development to provide economies of scale and enable the company to accelerate up the best practices learning curve in implementing data virtualization.

Corollaries to this need for centralized development responsibility are the following:

- Agree on and implement a common data model – The Global 50 Energy Company believes this will ensure consistent, high quality data, make business users more confident in the data and make IT staff more agile and productive.

- Establish a clearly defined information governance model – According to Qualcomm, this needs to include how to manage the data virtualization environment. Key issues are who is responsible for the shared infrastructure and for shared services.

- Ensure management support – Both the Global 50 Energy Company and Qualcomm described the importance of management support for the data virtualization effort. This may be easier if responsibility for data virtualization is centralized.

Educate the business on the benefits of data virtualization – Multiple organizations, including Northern Trust, Pfizer, the Fortune 50 Financial Services Firm, the Fortune 50 Computer Manufacturer and the Global 50 Energy Company emphasized the need to educate and support the business to successfully implement data virtualization. Ideas include:

- Allocate time to consult with business users and make sure they understand the data.

- Be prepared to provide support if the user has questions and diagnose problems accurately.

- Establish a culture of information sharing. Sharing and transparency are critical to increasing the value of the information.

- Internally market the concept of data virtualization. Establish an ongoing effort to make data virtualization acceptable in other areas of the organization. This involves educating people on its significant benefits and flexibility, taking a phased approach to expanding the scope of the data virtualization environment, and building incrementally on the success of each individual project.

- Expect resistance when bringing in a new approach/technology like data virtualization. There is a need to convince people who are more comfortable with a physical consolidation approach that data virtualization offers significant benefits and can solve a wide range of data integration problems.

• Manage business expectations. Implementation is faster but it still takes time.

> " *Multiple organizations ... emphasized the need to educate and support the business to successfully implement data virtualization.*

Pay attention to performance tuning and scalability – The case study organizations had clear advice here, including NYSE Euronext, Qualcomm, the Fortune 50 Computer Manufacturer, Comcast and the Global 50 Energy Company.

• Tune performance and test solution scalability early in the development process.

• Performance tuning expertise is critical. Ensure access to an expert in tuning the data virtualization suite/SQL to do the necessary performance tuning.

• Consider bringing in massively parallel processing (MPP) capability to handle query performance on high volume data and align the MPP and data virtualization implementations.

• Accommodate the fact that users are unpredictable on ad hoc analysis and reporting.

Performance of the data virtualization suite was a critical success factor for several of the case study organizations. Comcast, Qualcomm and the Fortune 50 Computer Manufacturer all cited the performance, reliability and scalability of their data virtualization solutions as key to their success. The Global 50 Energy Company emphasized its ability to achieve the required performance through the data storage component, stating that without the combination of data virtualization and MPP, its data virtualization solution would not have worked.

Take a phased approach to implementing data virtualization – Both Northern Trust and NYSE Euronext stressed the need to take a step-by-step approach to implementing data virtualization. It is important to first understand the data before coding views and attempting to integrate/federate the data.

NYSE Euronext counsels users to evolve the implementation step by step – first abstract the data sources, then layer the BI applications on top and gradually implement the more advanced federation capabilities of data virtualization. The Global 100 Financial Services Firm confirms that it is appropriate to start small with point implementations. This enables the organization to accelerate the business benefits and help fund larger deployments.

Qualcomm cautions user organizations to use data virtualization only where it is appropriate. Data virtualization is not a panacea or the solution to every problem.

Use an experienced vendor partner for data virtualization technology – Based on its experience, NYSE Euronext considers data virtualization one of the most complex things to get right, so do not attempt to reinvent this technology. The company advises other organizations to adopt an industry approach to data virtualization and partner with an experienced vendor with a mature product as an arm of IT to build out data virtualization capability.

Northern Trust suggests taking advantage of vendor professional services to help with difficult design and optimization challenges, avoid pitfalls and resolve issues quickly.

The Global 100 Financial Services Firm established close cooperation between the investment bank and the data virtualization vendor to design and fund necessary enhancements to the data virtualization suite. The company also stressed the importance of the quality of the people involved in the data virtualization effort. Both the Financial Services Firm and the data virtualization vendor committed their top architects and technologists.

Comcast and the Global 100 Financial Services Firm both cited the importance of the data virtualization's vendor's responsiveness and flexibility in enhancing its product to meet user requirements as critical success factors in their data virtualization implementations.

And finally, the representative of the Global 50 Energy Company offered this bottom-line advice: "Think outside the box to adopt data virtualization. Data virtualization works. Large software companies are on board. This is a direction in which enterprises should be heading."

Comcast

Organization Background

Comcast Corporation is a leading media, entertainment and communications company. Comcast operates cable systems through Comcast Cable, one of the largest video, high-speed Internet and phone providers to residential and business customers in the U.S. Comcast is also the majority owner and manager of NBCUniversal, which develops, produces and distributes entertainment, news, sports and other content for global audiences.

Comcast had approximately 22.9 million video customers, 16.7 million high-speed Internet customers and 8.4 million digital voice customers as of the end of 2010. Customers are located in 39 states and the District of Columbia.

The company was founded in 1963 and is headquartered in Philadelphia, Pennsylvania. There are approximately 102,000 employees nationwide. Annual revenue in 2010 was $37.9 billion.

We interviewed Craig Richards, Manager of Directory Services for this case study. Directory Services is responsible for developing and maintaining Comcast's provisioning data system for high-speed Internet and digital voice; developing web services to access the data using a service-oriented architecture (SOA); and providing enterprise-wide directory services for customer authentication and entitlements. The Directory Services team includes web service developers, LDAP (Light Directory Access Protocol) directory engineers and administrators, data delivery architects, analysts and data modelers. Directory Services is part of Comcast Enterprise Technologies (CET), Comcast's IT organization.

The Business Problem

Millions of Comcast customers use comcast.com or comcast.net as their portal to the Internet to access web sites and email and to manage their Comcast products and services online. Supporting this process involves many products and many disparate data sources that have to be integrated in a variety of ways depending on what the customer needs.

Two years ago, Comcast recognized the potential value of creating a federated, or virtualized, data architecture to integrate the data. The company conducted proof of concept tests on solutions from four vendors: Cisco, BEA Aqualogic (since acquired by Oracle and renamed), Radiant Logic and MetaData. Comcast selected Cisco based on the responsiveness of the support team, the fact that Cisco had most of the required features (such as flexibility and a web interface that could access multiple data systems), the company's strong track record and product pricing.

While Comcast saw broad applicability for the Cisco Data Virtualization Suite, the next step was to find a small application to get the product in the door and gain experience with it while solving a real business problem. The company decided to apply data virtualization to solve a key performance problem in the data source managed by Directory Services.

The Directory Services data source consists of two nodes, or data sets. The data comes in from a number of different sources through different web services and other interfaces.

- Global Directory Service (GDS) – GDS is a legacy system created about five years ago to manage subscriber provisioning data for high speed Internet and digital voice devices. Provisioning translates a customer order into actual service for that customer, enabling a physical device to provide the appropriate services (e.g., a phone line can get a dial tone).

- Enterprise Service Directory (ESD) – ESD was created as the repository for Comcast's customer identity and credentials data in its role as an Internet service provider (ISP). A user's identity connects the user with a Comcast account and information about who the owner of the account is. The user ID plus a password is the user's credential. ESD uses this information to authenticate the user; identify the permissions, or entitlements, if any, the user has to access email, manage the account, purchase new services, view digital TV content, etc.; and then authorize the user for those activities.

GDS and ESD are both LDAP directories implemented in CA Directory. In LDAP terms, each is a separate branch of the same schema and there is no built-in way, such as triggers, for the system to automatically maintain any relationships between the data in the two branches. Yet there is a business need to synchronize the two data sets because subscriber data on the provisioning side has a pointer, or reference, to the data on the identity/credential side, and ownership of an account is a key criterion for authorizing access.

The critical business challenge for Directory Services was optimizing the process of executing ownership changes on an account. According to Richards, "Ownership change is a complex problem to solve in real time. The customer is making what seems like a simple request to change the primary user ID on the account and doesn't realize the complexity involved. There are many things attached to your primary user ID, including a hierarchy of permissions. As the primary owner of the account, you can specify what your secondary user IDs can access. For example, if your child has a credential, you can control some of the content that child can see online. Changing the primary user ID means we have to switch all those permissions around. There is a whole layer of dependencies that we need to consider. In addition, there are legal requirements associated with changing ownership and/or credentials on an account."

To handle the cascading effects of an ownership change, Richards' group developed a custom schema and custom code for synchronizing (replicating) the data in the two directories. "Our configuration is extremely fast, much faster than any relational database could be. We use a combination of Linux and Solaris servers and all of the data is maintained in cache memory." The system is essentially an OLTP environment that can handle a peak load of 15,000 transactions per second. The average is about 25 percent of that. Ownership changes number a few thousand per day and the transactions are very complex.

While the directory was fast, performance on ownership changes was still a significant issue. First, it is a customer-facing application. The customer can request the change directly online or call a Comcast customer service agent who makes the change on the customer's behalf. In either case, the change has to occur very quickly to meet customer response time requirements. It was taking about five seconds on average and Richards said that his group could not figure out how to meet the mandated service-level-agreement (SLA) goal of two seconds or less with the current system.

Second, to execute an ownership change or update a credential, "we have to do the equivalent of multiple, iterative full table scans within the directory to collect all of the credentials and entitlements associated with the account. This clearly affects performance. We also have to synchronize the data between the two directories. But the system doesn't have the facility to do this in a single, atomic transaction to ensure consistency and accuracy across the two directories," stated Richards.

When response time is too slow or ownership changes are executed incorrectly, call volume goes up with customer complaints. If there are violations in the synchronization process, Directory Services has to spend time on extensive reconciliation activities to identify and correct inconsistent data between the two directories.

> *Data virtualization was brought in to improve both the performance and accuracy of ownership changes. The specific goals of the project were to provide change of ownership within the SLA requirement ... and to ensure the consistency and integrity of the data.*

Data virtualization was brought in to improve both the performance and accuracy of ownership changes. The specific goals of the project were to provide change of ownership for the customer and agent within the SLA requirement, and to do it within a single transaction to ensure the consistency and integrity of the data across the two data sets. Another requirement was the need to notify multiple downstream systems of the ownership change as well as comply with regulatory requirements for security purposes. While these notifications did not have to be synchronous with the ownership change transaction, it was important that they be managed as part of the ownership change process.

The Data Virtualization Solution

Comcast's entire ownership change process is now handled in a data virtualization layer managed by the Cisco Data Virtualization Suite (see Figure 1). This layer provides an abstraction of the two LDAP directory nodes (GDS and ESD) and all of the systems that must be notified when an ownership change occurs. The web services that execute the ownership change and notifications are built in Cisco as well. There is one service for the customer change and other services for notifications because it would take too much time for a single service to do all of the processing.

Here is a summary of how the process works. A customer requests an ownership change either online or by calling an agent on the phone. The customer or agent application connects to a web service hosted on Cisco. The web service issues a request for an ownership change to Cisco with the account number, the old user ID and the new user ID. Cisco first does all the processing to make the change in the directories in a single, atomic transaction. This includes all of the iterative logic required to modify the hierarchy of permissions within the identity dataset and to update both directory nodes. Cisco notifies the customer or agent application if the change is successful or not. Then Cisco invokes other services that take care of the necessary notifications to other sources and systems. These transactions don't have to be successful for the customer's ownership change to be successful. Notifications can happen separately.

Data sources are the two LDAP directory nodes. Data consumers are customer self-service applications, customer care applications, and the entitlement stores.

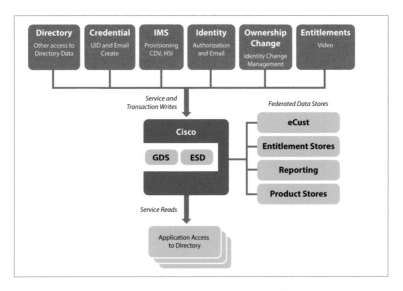

Figure 1. Comcast Directory Services Architecture for Ownership Changes
Source: Comcast

Richards stressed the advantages of having Cisco manage the whole process. The requesting service only has to make one request and Cisco then handles everything else that takes place for an ownership change using its native transaction management capabilities. "Using a pure web service would require multiple transactions in contrast to a single transaction within Cisco. A web service would also have difficulty with the iterative nature of the processing and managing the multiple full data scans needed within the directory to match up the data."

Comcast has more than achieved its SLA response time goal of two seconds. "Our average response time with Cisco is now 1.2 seconds, better than we expected. In fact, it's pretty darn good. We were hoping to get under 5 seconds," said Richards. He added that Cisco helps make the process fast by managing the iterations that take place, updating both sides of directory synchronously and managing the notification process.

The Implementation Process

The data virtualization project began in early 2009 and the system went into production in late 2009. Implementation took longer than originally anticipated for three primary reasons. One was that the solution requirements changed while the Directory Services group was building the system. Another was the fact that Cisco needed to add support for an LDAP data source environment to the Cisco Data Virtualization Suite. Comcast contracted with Cisco to create the LDAP translation, and once that was done, Richards' group could proceed with building the web services for handling ownership changes. Here, according to Richards, Comcast encountered some additional issues. "Our SOA standards weren't compatible with the way Cisco interfaced with web services, specifically in the area of the Web Services Definition Language (WSDL) and publishing of the WSDL. Cisco had to make some changes to fit into our SOA framework and did what we needed."

Summary of Benefits and Return on Investment

BENEFITS. Richards described several significant benefits Comcast has achieved with its initial data virtualization implementation.

Improved customer satisfaction and self-sufficiency – By reducing the time needed to process a customer request for an ownership change from as long as 10 seconds down to 1.2 seconds, Directory Services has been able to easily meet the required SLA of 2 seconds or less. Another aspect of meeting the SLA is the fact that Cisco has been consistently available. Richards stated that Comcast has never experienced an outage with Cisco so far. "The customer always can reach the system. Cisco has been extremely reliable for us."

> *By reducing the time needed to process a customer request for an ownership change from as long as 10 seconds down to 1.2 seconds, Directory Services has been able to easily meet the required SLA of 2 seconds or less.*

Giving customers fast response time and a successful outcome when they contact Comcast online makes them more likely to use a self-service approach in the future and minimizes the potential for unsatisfied customers. This also contributes to lower call volumes.

Reduced cost of development and maintenance – Transaction management for ownership changes has been pushed down into the data virtualization layer from the application/services layer to take advantage of native Cisco functionality. This simplifies application development and maintenance. It also eliminates the need for complex reconciliation processes across the directory data sets when they get out of synch.

Easier access to LDAP data – Another benefit mentioned by Richards is the fact that the data virtualization layer now provides the capability to execute SQL queries against an LDAP database. "There are people who want access to the LDAP data but have no clue how to talk to LDAP. In contrast, a lot of people are trained in writing SQL queries. Providing an access layer to LDAP for people who know how to write SQL makes LDAP less intimidating and increases our ability to make additional use of the information." Examples are applications that analyze churn rates or the extent to which customers are using specific products. "We are using Cisco in a limited way to access LDAP today, and there are a lot more opportunities out there."

Richards summed up by saying, "We really like the way Cisco works with us as a customer. One reason that its solution won out was the company's responsiveness. When we had problems, they made it a top priority to solve those problems. That's what you want from a vendor partner."

RETURN ON INVESTMENT. In terms of ROI, one example Richards cited is the cost savings that occur when customers use a self-service application to make an ownership change instead of calling customer service. He estimated that Comcast saves about $2,000 per day by making customers more self-sufficient.

Summary of Critical Success Factors

Richards cited two major factors that have contributed to the success of the data virtualization solution. "The two big factors that are making this system work for us are performance and the consistent uptime and dependability of Cisco."

Future Directions

One future use of data virtualization, discussed in the "Summary of Benefits" section above, is to allow SQL queries to access the LDAP directory data. Richards also wants to extend the use of data virtualization throughout Comcast. In his opinion, "A data virtualization solution is much easier and more streamlined than creating a data warehouse and using ETL tools to synch everything up." His group is trying to get other departments to understand the value of the data virtualization architecture and recognize that it will solve a lot of the problems they have. At the same time, he acknowledges that it is "a hard sell. This is a new approach and people are often more comfortable dealing with what they already know."

Compassion International

Organization Background

Compassion International is the world's largest Christian child development and sponsorship organization. Its mission is to permanently release children from spiritual, economic, social and physical poverty and enable them to become responsible and fulfilled Christian adults. Compassion has three programs in place to accomplish this.

The Child Sponsorship Program educates children on core issues such as health, nutrition, life skills training and the Bible to help end the cycle of poverty. A key component is matching a sponsor with a child in need.

The Leadership Development Program identifies Compassion-assisted children who have exhibited leadership ability within their communities and helps to further mold them through a college education. These children have illustrated the drive and determination to affect positive change in their communities but lack the resources to do so.

The Child Survival Program enables Compassion to start care and intervention even before a child is born. This program targets prenatal mothers and helps children between the ages of 0 and 4 get a healthier start on life.

Because poverty is a very complex issue, Compassion partners with over 5,400 local churches around the world to deliver services. The local church can best identify and handle the specific needs of children in its area. Compassion's role is to partner with and equip the local church and provide sponsor support for the children it serves. Compassion also maintains partner offices in 12 countries throughout the world.

Compassion was founded in 1952 by Reverend Everett Swanson to provide Korean War orphans with food, shelter, education, health care and Christian training. Today, the organization helps over 1.2 million children in 26 countries with a vision to have 4 million beneficiaries by 2020. Over 80 percent of Compassion's expenditures go directly to program activities. The organization is headquartered in Colorado Springs, Colorado.

For this case study, we interviewed Kenny Sargent, Technical Lead, Enterprise Data Products. Sargent's team is the technology arm of Compassion's Enterprise Information Management (EIM) group and is staffed by nine people with another two for the EIM group. Compassion's overall IT team is staffed by 150 people.

The Business Problem

In the 2007/2008 timeframe, Compassion recognized the need to modernize its information systems if it expected to achieve its goal of quadrupling the number of children served over the next decade. The systems in place at the time could barely meet existing information requirements, let alone carry the organization successfully into the future. Compassion clearly needed a new information infrastructure that would enable the IT team to achieve a number of specific objectives:

- Respond more quickly to new end-user requests for information.

- Support Web 2.0 collaboration applications to improve the overall flow of information throughout the organization for both internal and external users and facilitate more effective communication.

- Reduce the total cost of application development, ongoing maintenance and system administration and free up resources for Compassion's core charitable mission.

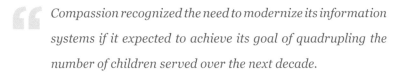 *Compassion recognized the need to modernize its information systems if it expected to achieve its goal of quadrupling the number of children served over the next decade.*

The resulting initiative was called the Capacity Building Initiative (CBI). The enterprise data solution, a subset of CBI, involved the complex challenge of implementing organizational changes and new technology while ensuring that business processes were both disciplined and flexible enough to adapt to changing requirements. Goals included:

- Create an Enterprise Information Management (EIM) program and supporting organization to ensure that information is treated as a corporate asset and needs are evaluated and addressed at an enterprise level. EIM would be a dedicated, non-IT business information and governance group with responsibility for addressing taxonomy and consistent business rule application throughout the organization.

- Implement the required technology support for EIM, including a corporate information factory, enterprise data warehouse and master data management (MDM).

- Establish a comprehensive data governance program, administered by EIM, that involves the business side of the organization as a partner with IT at all levels – strategic, tactical, and operational.

A key requirement on the technology side was flexibility and agility in integrating data sources to support reporting, analysis and decision-making across the enterprise as well as the EIM effort. Compassion decided to implement a combination of data virtualization, enterprise data warehousing, ETL and MDM technologies to accomplish this. In 2008, this seemed like the best of breed solution set based on the technologies available at the time. By late 2010, it became apparent that the optimal solution was an architecture that greatly leveraged data virtualization.

The Data Virtualization Solution

Compassion's data integration solution is the Ministry Information Library (MIL) developed and managed by Sargent's team within IT. The MIL is Compassion's implementation of a Data Delivery Platform[1] that provides an enterprise-wide, single version of the truth to answer questions and provide corporate alignment of numbers and metrics as well as definitions of core business entities.

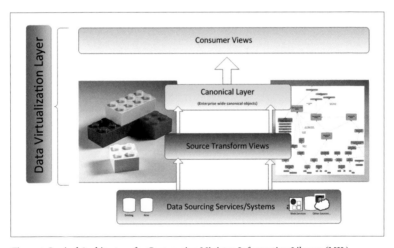

Figure 1. Logical Architecture for Compassion Ministry Information Library (MIL)
Source: Compassion International

The MIL provides read-only access to much of the organization's data for both internal and external users, and includes a data virtualization layer implemented with the Cisco Data Virtualization Suite. The logical architecture of the MIL is shown in Figure 1.

---

[1] *A Definition of the Data Delivery Platform by Rick van der Lans, BeyeNETWORK, January 21, 2010 (http://www.b-eye-network.com/channels/1134/view/12495)*

Data sources – Data sources include internally-hosted application repositories (e.g., ERP, CRM and financial data) and can be repositories hosted in the cloud (e.g., an external CRM repository that supports relationship tracking and market analysis). An additional source is Compassion's Enterprise Data Warehouse (EDW). Data from these repositories is either accessed directly from the data virtualization layer or extracted for the MIL using ETL tools depending on the service-level agreements (SLAs) with the source-system owners.

Data virtualization layer – The data virtualization layer is a set of views that integrate data across various source systems or source system copies, virtually transform data for enterprise agreement and deliver the data to consuming applications. Here, data virtualization provides significant benefits.

- Hides complexity from data consumers by abstracting and federating the source data or source data copies, making data sources independent of data consumers. This means Compassion can swap out a data source without having to make any changes to the application layer. This enables IT to establish what it calls "fixed contracts" with data consumers (i.e., to deliver specific data in a specific format) while retaining the flexibility to change the underlying data source and/or the logic of how the data is actually provisioned.

- Allows EIM and the MIL to facilitate business agreement through canonical views that have semantic, enterprise agreement on terms and rules that then surface data in very consistent ways, with almost no use of ETL. As Sargent said, "The data virtualization layer combines good, old-fashioned applied set theory with some modern tools and architectural methods to facilitate enterprise agreement, allow for significant reuse and provide a responsiveness to customer requests that has been unparalleled in our experience." According to Sargent, internal customers have exclaimed, "Wow, this is amazing."

- Allows the same set of agreed-upon business rules to be applied consistently across integration techniques to improve information quality.

- Allows the MIL to surface data at any level of latency required by the customers from near-real time to daily/monthly/yearly.

Consuming applications – These include portals, reporting and analysis tools and other applications. Examples are Excel, PerformancePoint, SAS and SPSS. In addition, Compassion's EDW also consumes data from the data virtualization layer.

FLEXIBLE DEPLOYMENT. The use of data virtualization within the MIL has enabled Compassion to meet demanding new information requirements much more easily and quickly than before. Here are two examples.

In 2009, Compassion designed a new, Facebook-like, social networking site called ourCompassion.org for sponsors and supporters of children living in poverty. This was to be a significant global effort and the data was to be delivered through the MIL and Cisco. This included posting on the site photos of each of the one million children registered in Compassion's programs. Just before officially launching the site, Compassion discovered that all one million photos were corrupted and unusable. The organization was able to avoid delaying the rollout by simply redirecting Cisco to access the original file of photos. No other changes were necessary. This illustrates the flexibility data virtualization provides by making data sources independent of data consumers.

In another case, Compassion's market research group wanted to assess whether letting users access its web site and execute financial transactions without forcing the user to first log in to his/her account would increase the amount of income flowing through the site. Sargent's team constructed a service in Cisco that took in massive amounts of XML data each day, staged the data in a cache and made it available to a SAS data mining application for analysis. Design, construction and deployment of the service took only a matter of days with data virtualization instead of weeks or even months. This highlights how the ease of developing a new web service and adding a new data source within the Cisco environment can reduce IT's time to solution for new application requests.

VALIDATION-DRIVEN ENTERPRISE INFORMATION MODELING. Sargent's team also developed a semantic, or definitional, layer for the data as the very first step in creating the data virtualization layer. The key was to establish and agree on the language semantics that matched the business definitions and business rules and embed them in a set of virtualized views. These views would then represent how the business wanted to define and report the data – that is, the business definition of a "sponsored child," a "registered child," a "sponsorship," etc.

The views would represent a hierarchical set of data building blocks that could be standardized and layered on top of each other to surface source system data, define canonicals and put together meaningful blocks of business functionality for read-only purposes. This ties back to the objective of the EIM organization to apply taxonomy and business rules consistently throughout the organization. Sargent summarized the process of validation-driven development as follows.

- Establish a quick, stake-in-the-ground, first cut at the business definitions (mutually exclusive states or buckets of data).

- Define a first cut at the business rule equations necessary to validate the definitions. The goal of these first two steps is to get at the data as rapidly as possible.

- Write SQL to do the validation. This is part of a rapid-prototyping effort in Cisco and SQL because the validation SQL and the Cisco building blocks go hand in hand.

- Present the results of the rapid prototype back to users and have a conversation around, "This is what we found. Does this make sense?" This "fail faster" process allows Compassion to have meaningful discussions early in the process and deal with the biggest risks at the earliest points in the process.

- After the process has iterated such that the above issues have been sufficiently worked out, the team "hardens" the development versions of the Cisco views into building blocks that are reusable. There are different building blocks at different layers, but the goal is always reuse.

Here are the characteristics of the building blocks.

- Each building block is an actual view that can be used and queried, not just a logical object.

- From the bottom up, the building blocks (views) get more business friendly and less like a view of the source system. The middle blocks in the diagram represent agreed upon canonical objects.

- The building blocks standardize the business logic so the same logic doesn't have to be duplicated across the organization.

- Different building blocks or layers can be physically cached for performance and/or controlled latency (see below) if necessary.

- Each building block is intended to have a link to an internal wiki page to document it. These are published and accessible to end users and developers.

Sargent considers this effort still a work in progress. "The heart of our information architecture is to push the semantic agreement upstream. The agreement is in our data virtualization layer as views. They are still physical objects that can be used, but they are lightweight and easy to modify. The goal is to create a set of 'legos' that can be put together and reused. Caching for performance and actual storage of data happens downstream from the views, and only if necessary."

CONTROLLED LATENCY. An important feature of the data virtualization solution is caching for what Sargent calls "controlled latency." This is the ability to offer the user different options for data latency and, therefore, expected performance on queries. Sargent's group creates cached snapshots of the MIL data once a month for "canonical counts" and these are persisted in the EDW. Cisco also caches snapshots of the data on a daily basis if desired. The data virtualization layer allows for any level of latency including near-real time, hourly, daily, weekly, monthly, quarterly, yearly and various combinations of those. For example, there is one business need to combine up-to-the-minute and historical data. Using data virtualization to federate the near-real time and historical data has allowed the business need to be met with very acceptable performance. Typically, near-real time is used for small, up-to-date queries such as "get me the latest on this child," and the performance is usually sub-second.

The Implementation Process

Compassion commenced its data virtualization implementation in 2009, focusing first on flexible deployment of data virtualization to enable individual application solutions. Over time, the organization has extended its data virtualization implementation on a number of dimensions to cover more applications as well as to address broader reuse by expanding the validation-driven enterprise information modeling approach.

The more the team has understood Cisco's query optimizer and how to architecturally leverage controlled latency, the more it has been able to deliver the right data, on the right schedule, to the right users with acceptable performance.

Summary of Benefits and Return on Investment

BENEFITS. Sargent described a number of significant benefits realized from the data virtualization solution.

Improved data quality and integrity – This has ramifications far beyond the ability to deliver consistent results on reports to users of the MIL. Data integrity is critical to Compassion's ability to continue the good stewardship of its mission and practice. "We care about every single child in our programs and we want to make sure the story we tell is supported by our data. Our validation-driven development approach built on the Cisco suite is helping us to do that. It is also important to be able to show any organization that evaluates charities that we are spot on with our data integrity."

Faster turnaround on new requests for data – Data virtualization gives IT a flexible foundation on which to quickly implement new data solutions and to develop services that are reusable across the enterprise. As a result, the MIL team has reduced the time it takes to respond to new end-user requests by more than 50%. As mentioned above, the team has been able to "fail" much faster.

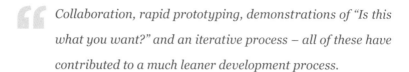

> *Collaboration, rapid prototyping, demonstrations of "Is this what you want?" and an iterative process – all of these have contributed to a much leaner development process.*

In addition, they are able to deal with the biggest risks (issues with the data) much earlier in the process. Furthermore, a process for quickly getting to signoff has been emerging. Collaboration, rapid prototyping, demonstrations of "Is this what you want?" and an iterative process – all of these have contributed to a much leaner development process.

Reduced IT cost for high-quality data – The total cost of developing data services and reporting and analytic applications has been reduced by at least 25 to 35% per project. This has allowed IT to do more projects and make more useful information available to the ministry. As part of a recent win, one internal customer said, "Thanks for taking the time to explain [this] and its benefits! This tool is so much more accurate and detailed than [our previous reports] and takes all the research out of the child pool and longest waiting children. I can see incredible uses for this data and am looking forward to discovering how to fully utilize this information."

RETURN ON INVESTMENT. While Compassion has not yet identified an overall ROI for the MIL, it has achieved measurable results in several areas as described above.

Summary of Critical Success Factors

There were two factors critical to Compassion's success in implementing the MIL data integration solution.

The flexibility of data virtualization – Here are some examples of how this has contributed to Compassion's success in developing the MIL.

- Data virtualization simplifies data integration, making IT more responsive and effective in meeting new requests.

- Data virtualization enables all data sources to act as a single source, making it easier to deliver data in multiple ways for different consumers.

- Data virtualization enables Compassion to modify things without affecting the end-users' experience.

- The ability to standardize enterprise business logic and business terms with views instead of ETL simplifies overall maintenance and makes it much easier to develop new solutions.

A maturing EIM organization – Compassion's IT and overall leadership had the foresight to put in place a bona-fide, EIM group vested with executive authority. This has been a significant help in facilitating cross-functional agreement. The organization is maturing and providing real leadership value in defining Compassion's information architecture.

Sargent measures success based on a simple metric: are the customers really happy and are we moving in the direction of true enterprise agreement? Until late 2010, even though the MIL was garnering industry attention, he was honestly not feeling that the customers were very excited. Then, when he was at his lowest point in terms of believing that this collaboration could work, he said, "God opened a significant door for us in our understanding, showed us a new architecture and provided some very key strategic relationships." The subsequent successes over the next few months were born out of that.

Future Directions

Much of the future direction is continuing and maturing what Compassion is doing now. Sargent believes very strongly in continuous improvement and lean principles. He wants to identify the true value-added activities and maximize them while minimizing and hopefully eliminating no-value-add activities. In addition, he says, "we're having fun again."

Fortune 50 Computer Manufacturer

Organization Background

This data virtualization user is a multinational company that develops, manufactures, sells and supports computers and computer-related products and services worldwide. The company ships over one million systems per week. Including orders for software, service and peripherals, there are over 5 million orders in process at any time. To support this, the company averages two billion customer interactions per year and has a supply chain in excess of $40 billion annually.

For this case study, we interviewed an Enterprise Architect in the organization's supply chain IT area.

The Business Problem

A few years ago, following a change in strategy, the company decided to outsource much of the design, manufacturing and build/assembly of its computers to third-party partners known as ODMs (original design manufacturer). Prior to that, the company manufactured its own computers, handling internally the entire process from customer order to manufacturing to delivery. This was done on a regional basis (e.g., Americas, Europe/Middle East, Asia Pacific), with each region responsible for fulfilling and tracking its own customer orders. The company described itself as global, but in fact operated as separate independent regions with each region responsible for its own operations, including supply, inventory, manufacturing and logistics.

Currently, ODMs handle over half of the company's manufacturing and this percent will likely continue to grow in the future. The decision to outsource much of the supply chain also forced the company's IT department to take a more global approach to its supply chain data strategy as orders from multiple regions could now flow globally into one or more ODM facilities worldwide. For example, orders originating from American and European customers might now flow to the same ODM factory in Asia. "The regional separation of IT systems didn't make sense anymore. We needed a global view of operational data and a system that could communicate sales orders and 'build' instructions to our ODMs and get messages and data back from them in near-real-time. The key was to take a data-centric rather than an application-centric view of the overall process and data," stated the Architect.

There are four primary functions in the company's supply chain management system:

- Planning translates quarterly revenue forecasts into volumes of specific products within each business unit that need to be produced in each ODM facility each week. This information is then used to determine, via a bill of materials (or "BOM") explosion what parts and how many are needed in each facility each week.

- Procurement takes the resulting "shopping list" and works with global suppliers and shippers to order the parts and ensure they arrive where and when needed. This complex inventory management process involves thousands of parts and hundreds of suppliers and facilities.

- Order processing receives orders from customers and routes them to the appropriate ODM for fulfillment.

- Logistics then routes the final products to the end customers.

A critical component of creating a global view of supply chain data was globalizing the procurement reporting system. This presented several challenges. The company used a third-party procurement application and there were six separate regional instances, each with its own repository of data that tracked inventory positions within that region. So the first requirement was to provide an integrated, global view of the procurement data across the six instances.

The second challenge was to implement a reporting system that could provide users with flexible access to and analysis of integrated data. To get information about a global part position (e.g., how many 500GB hard drives are available, what is on order, etc.) in the existing environment, a buyer had to request six separate reports, one from each regional inventory instance, then pull the data into an Excel spreadsheet and manually combine it. This was time-consuming, error prone, and made it very difficult for buyers to effectively analyze data and make timely decisions (e.g., adjusting the supply of parts when the company is over or under forecast).

A third challenge was to ensure that a new reporting system did not introduce any latency. The business would only accept a solution that could deliver the same up-to-date information that was available through the current procurement applications.

The fourth challenge was the need to replace a legacy reporting environment running on an outdated operating system that posed security risks.

The team evaluated two alternative approaches to globalizing the procurement data and reporting system.

- Globalize the data physically – Create a single global database and use traditional ETL tools to copy the data from each regional instance into an integrated data warehouse. According to the

Architect, the company had already taken this approach in some areas, creating an operational data mart which consolidated global order data into a single physical database. Physical integration with ETL, however, "caused its own set of problems, including data latency and redundant copies of data and redundant storage requirements which, in turn, increased our infrastructure costs."

• Globalize the data "on the fly" through data virtualization – Create virtual views to federate the data in the six repositories and deliver global, integrated data through these views to the reporting system. This approach had distinct advantages. It avoided the need to replicate any data, reducing infrastructure cost, and did not require any changes to the procurement application environment, reducing implementation complexity and time. In addition, data virtualization did not introduce any latency.

Supply chain IT chose the data virtualization approach and evaluated both the Cisco Data Virtualization Suite and BEA Aqualogic (since acquired by Oracle and renamed). The team selected Cisco because of its performance and support for SQL. "SQL is a familiar tool for us and the existing code in the procurement system is all in SQL," explained the Architect. In addition, Cisco was a data virtualization specialist with a mature product offering.

 Data virtualization had distinct advantages. It avoided the need to replicate any data, reducing infrastructure cost, and did not require any changes to the procurement application environment, reducing implementation complexity and time. In addition it did not introduce any latency.

The Data Virtualization Solution

The company's solution is a procurement reporting system that combines a third-party BI reporting tool supported by data virtualization to deliver a global view of inventory data to procurement users. The architecture is shown in Figure 1. The

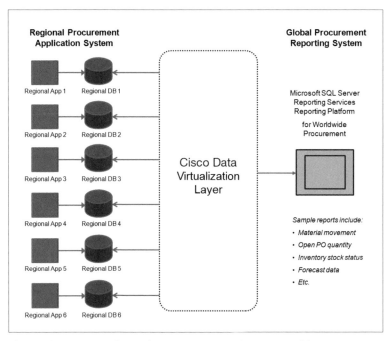

Figure 1. Computer Manufacturer's Procurement Reporting System Architecture
Source: *Fortune 50 Computer Manufacturer*

solution was implemented without any changes to the existing procurement application or the underlying transaction systems.

Data sources – The data sources are the six regional instances of the procurement data. Each contains multiple terabytes of inventory and material supply data and hundreds of individual application tables.

Data virtualization layer – The data virtualization layer globalizes the source data through a set of views that directly access all six regional repositories and federate the data for delivery to the BI reporting system. Any required transformations of the source data are handled within the data virtualization layer through the views. Because this layer queries the same source data as the procurement application, data virtualization does not introduce any data latency. Users see the same data whether they are accessing the data through the reporting system or through the procurement application.

Consuming application – Microsoft SQL Server Reporting Services (SSRS) is the BI platform that generates the procurement reports. The supply chain IT group created over 100 new reports to replace the existing legacy procurement reports and Excel spreadsheets. These reports support approximately 300 global procurement users, including buyers and inventory control analysts. Users can now easily access and analyze global procurement data with the new system – open purchase orders, supplier quantities, material supply lines, global inventory levels, etc. On a daily basis, procurement users now execute thousands of queries and generate hundreds of reports, sourcing millions of rows of data through Cisco.

The Implementation Process

The company decided three or four years ago to evaluate data virtualization as a potential data integration solution. The company issued an RFP, talked to several vendors and selected Cisco as the one that best fit the requirements. Cisco was originally brought in to address specific application needs in the supply chain area, such as the procurement reporting system. Cisco has since become an enterprise tool in the enterprise data integration area of IT.

When the company decided to globalize its procurement data using data virtualization, it also decided to migrate both the existing procurement system reports and the legacy reporting system to the SSRS platform. Creating a unified reporting environment became yet another valuable opportunity to leverage the unified data globalization effort.

To implement the procurement reporting system, the company created a team consisting of both internal and external resources. Internal IT staff developed the Cisco views because IT understood the business and the data and was experienced with Cisco from earlier projects. This effort required less than two months. The reports were developed by outside contractors. Rewriting the reports and converting them to SSRS required three to four months.

This team approach reduced overall development time and enabled IT to deliver the solution to the business faster.

ADVICE. The Architect offered two pieces of advice to other organizations. One is that, even for a solution used elsewhere in the company, it is important to expect resistance when bringing in new technology approach like data virtualization. "We had to convince the company that the approach of globalizing data on the fly using data virtualization was better than the physical consolidation approach with which people were more comfortable."

He also cited the importance of tuning performance and testing the solution's scalability early in the development process. "We had used Cisco for a number of years under fairly light work loads. When we started moving hundreds of procurement reports over to Cisco, we went from a light to a heavy load very quickly. We discovered some performance problems that didn't show up until the users were on the system. These problems were subsequently resolved via query tuning and the intelligent use of caching but could have been uncovered sooner."

Summary of Benefits and Return on Investment

BENEFITS. The Architect described several major benefits the company achieved with its data virtualization solution.

An optimal solution – "We implemented the desired globalization at both the BI and data levels while continuing to maintain our regional application instances. This was important for autonomy, performance and legacy reasons." Data virtualization also enabled the company to provide direct access to the current source data with no added latency.

Reduced cost – On the cost side, data virtualization helped the company avoid the infrastructure costs associated with physical integration in terms of redundant data and storage. In addition, Cisco's support for SQL enabled the supply chain IT team to maintain high productivity and minimize staff disruption.

Faster time to solution – With data virtualization, the supply chain IT group was able to quickly deliver to the business a solution that provided a more flexible reporting system and globalized data. As the Architect stated, "Cisco helped us extend our infrastructure in a quick and painless way because it fits nicely into our existing infrastructure. It is also SQL-based and can access our existing procurement application databases without requiring any changes to either the application or the data sources."

 Millions of dollars per year are saved through faster inventory turns and improved customer satisfaction. The ability to deliver the solution faster using data virtualization also saved over a million dollars in development and infrastructure costs.

RETURN ON INVESTMENT. The company has successfully demonstrated significant ROI from its procurement data virtualization and reporting solution. Millions of dollars per year are saved through faster inventory turns and improved customer satisfaction. The ability to deliver the solution faster using data virtualization also saved over a million dollars in development and infrastructure costs.

The Architect added two perspectives here. One is that migrating off the legacy reporting system enabled the company to eliminate potential security risks and reduce the cost of ongoing maintenance. Another is that the globalization effort is paying back in a cumulative, incremental way. "Every time a global buyer runs a report, we accumulate some additional return on our investment in the data virtualization solution."

Summary of Critical Success Factors

The Architect described three aspects of the procurement reporting solution that were critical to the company's success. One was the scalability and performance of the Cisco solution.

Another was Cisco's support for SQL. This made Cisco very easy to use for report developers and database programmers. "Cisco was a good fit for our team and our infrastructure and did not require a steep learning curve."

A third key to success was the fact that the company did not have to make any changes to the source data when implementing the procurement reporting system. "The transformations we needed to make, which were very minor, were all handled in the Cisco layer through the views," stated the Architect.

Future Directions

The Architect would like to see data virtualization used more broadly as an enterprise solution but recognizes that "we need to do a better job of promoting the benefits of data virtualization internally."

One area of future potential is to help the company integrate operational (near-real-time) and data warehouse (historical) data. "We tend to think of these as totally separate today, but there are opportunities to bring them together." One example is global order visibility, or the ability to combine live orders in operational systems and historical orders in the data warehouse through a common virtualization layer to enable the user to see any order, regardless of where it resides. The issue here is that the current tracking system for live orders (over 5 million in process at any time) is "just about as big and as fast as we can make it now." The company is about to upgrade the database system but it can only hold live orders and these have to be purged quickly to free up space. With data virtualization, the order search queries could be directed first to the live order database, then to the data warehouse if the order is historical.

Fortune 50 Financial Services Firm

Organization Background

This data virtualization user is a Fortune 50 financial services firm headquartered in the U.S. The company offers a full range of financial services for retail consumers and corporate and commercial customers through thousands of branches and other worldwide distribution channels.

For this case study, we interviewed the Vice President responsible for designing and developing solutions for the wholesale banking function.

The Business Problem

This firm faced a major business challenge when it acquired another large financial services organization a few years ago. Melding two large companies is never easy but this merger was completed under particularly difficult circumstances. It took place in the aftermath of the 2008 financial crisis along with significant changes to financial regulations. The Vice President described the result of the latter as a "post-friendly" regulatory environment in which financial regulations became much more important, demanding and introspective compared to those in place in the pre-2008 timeframe.

The most urgent concern in completing a merger is the integration of systems and data across the two companies. Depending on the size of the merger, instead of just two systems that perform the same function, it is not unusual to see four or five if one or both companies are themselves the result of previous mergers in which multiple similar systems have been cobbled together. Examples in financial services are retail customer account data (checking, savings, etc.), credit card data and systems that capture trading data and manage customer investment portfolios. The process of decommissioning redundant systems and moving all of the data into a single enterprise system becomes a massive integration effort that must be repeated for every different type of core operating function and line of business.

 The most urgent concern in completing a merger is the integration of systems and data across the two companies. Depending on the size of the merger, instead of just two systems that perform the same function, it is not unusual to see four or five if one or both companies are themselves the result of previous mergers.

Solving this integration problem, according to the Vice President, demonstrates "our appetite for ETL-style integration technologies. We often find ourselves almost blindly going in this direction, given the tight deadlines to merge operations, without thinking about whether moving and physically consolidating the data actually makes sense. As a result, most integration projects become ETL projects. If you are in the middle of a merger and you don't know any other solution than to move the data, you move the data." Yet the process of moving data using ETL technologies can be more costly and riskier than allowing an existing, stable data source to remain as is – as an isolated line-of-business (LOB) data store – while providing enterprise-wide access to the data through virtual views and services on top of the data. These virtual views may also make it possible to decommission the satellite systems that surround the LOB data except for applications that directly manage the data.

The non-technical aspects of ETL integration are also important to consider, including the time constraints and cost of buying new infrastructure and provisioning it. It may take months for a new production system to arrive for managing consolidated data, plus additional time to get it up and running, create the integrated/ transformed data structures and instantiate the data. A company may not have the luxury of waiting that long to begin integrating the data.

Instead, the Vice President stated, "We took a step back, breathed easy for a moment and assessed how many 'automatic' ETL projects we could potentially avoid if everyone were educated on the benefits of data virtualization. Data virtualization provides a single point of data consolidation and a single interface for delivering data across diverse sources to multiple consumers. We decided to be creative and aggressive in our merger effort to implement rapid data integration using data virtualization technology."

The Data Virtualization Solution

This firm's data virtualization solution spans three phases: pre-merger, merger and post-merger.

PRE-MERGER. Prior to the merger, the acquired company had already embarked on a number of successful data virtualization projects using the Cisco Data Virtualization Suite. These efforts resulted in a data virtualization layer that integrated access to over 200 disparate data sources for more than 25 applications serving approximately 500 data consumers (see Figure 1). The company had two goals in implementing its data virtualization solution.

The first goal was to provide a more flexible and lower-cost infrastructure for data access by abstracting the data and decoupling the data sources from data consumers. Benefits include the ability to change data sources without requiring any changes to the consuming applications and to reduce infrastructure and development costs.

The second goal was to increase business opportunities by using data virtualization to create an API, or service, that made it easier and faster for a new application to access existing data and functionality. Examples are exposing a virtual view as a web service and/or a JDBC/ODBC interface. There are several benefits here. First, the developer doesn't have to spend the time and effort to write ETL routines on the "data in" side, or write wrapper web services and figure out where to host the wrapper on the "data out" side. Data virtualization greatly simplifies and standardizes all of this complexity because it is all developed in and hosted within the data virtualization suite.

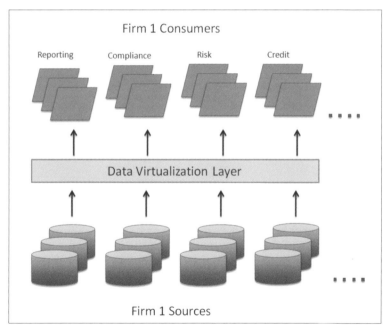

Figure 1. Acquired Company's Pre-merger Data Virtualization Architecture
Source: Fortune 50 Financial Services Firm

Building an API/service also makes the data and functionality available for delivery and use in multiple form factors to support different devices, interface styles, navigation methods, information requirements, etc. The key point here from the Vice President's perspective is that "it is more important to build a business API to access the data than to build the business application itself. When you build a business application, you have already decided on the form factor, the target device, the navigation style." The alternative approach of delivering an API offers far more flexibility for customers and more revenue opportunities for the firm. The data and how it is used and presented can now be customized, integrated, rebranded, and/or embedded in other applications, services and systems to meet unique business needs. Here, Cisco's unique ability to quickly discover data and build out an associated services layer for accessing data is immensely valuable."

> *We discovered we could also apply data virtualization to rapid integration of our data and minimize the need for more costly, less efficient ETL and data-movement projects.*

MERGER. Through its successful experience with data virtualization, the acquired company had firmly established that this approach could be used to provide more flexible access to existing data and enhanced overall agility without needing to change the underlying data sources. Therefore, when the merger opportunity arose, the idea of using data virtualization to quickly integrate data during a merger/acquisition process became a logical extension of the original use case. According to the Vice President, "We discovered we could also apply data virtualization to rapid integration of our data and minimize the need for more costly, less efficient ETL and data-movement projects." This involved creating virtual views of data across the two merged companies to give data consumers a single, integrated view of the new entity without having to move the data first (see Figure 2).

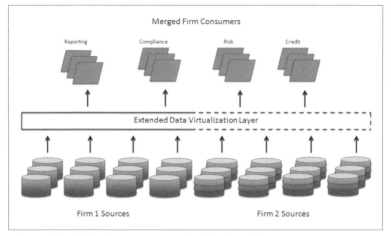

Figure 2. Financial Services Merger Data Virtualization Architecture
Source: Fortune 50 Financial Services Firm

POST-MERGER. In the post-merger phase, the focus shifted back to the ability to implement a technology and business "refresh" by creating services to simplify and improve access to data and application functionality. This now involves looking at new ways to deliver the same functionality more efficiently and to increase revenue opportunities for the combined companies. As the Vice President put it, "During the merger, we are moving at 200 miles per hour and trying to do things very quickly. Unfortunately, we often do them inefficiently and spend more than necessary." An example for this organization is the fact that each party in the merger had its own commercial banking portal. The merged organization needed one portal with the branding of the acquiring company. Once the initial integration between the portals is accomplished through virtual views and services, the company can go back and change, or "refresh," the underlying technologies and the way the application is delivered if appropriate without affecting the data consumers on the front end. "To do this, we used the same techniques we learned from our efforts before the merger."

The Implementation Process

In this case, as described above, the acquired company had experience using data virtualization for a variety of projects within its investment banking division since early 2006. One of the primary use cases was to provide client, holdings and trades data to multiple trading and client management applications. This project was completed in less than six months. During the merger process, the companies extended this use of data virtualization to quickly integrate data across similar functions and lines of business without having to change the underlying systems.

ADVICE. The Vice President acknowledged that using data virtualization may not be politically viable for some organizations. In spite of the increasing penetration of this approach within his firm, "we have experienced several missed opportunities." In fact, he said data virtualization was, in many cases, a "tough sell. Many people are trained to do things in one particular way. For data integration, it is ETL technologies and moving data."

The key to success, stated the Vice President, "is educating people on the significant benefits and flexibility of data virtualization. It is teaching them a new trick." While the technology is well accepted in one division of the firm, there is an ongoing effort to make a data virtualization approach acceptable within other divisions as well. This involves not only education as one critical implementation step, but also the use of a phased approach to expand the scope of the data virtualization environment by building incrementally on the success of each individual project.

Summary of Benefits and Return on Investment

BENEFITS. The company has achieved several major benefits from its data virtualization solutions.

Ability to meet critical time constraints for integrating data during the merger – Data virtualization enabled the company to immediately integrate data in areas where it was important to present a seamless transition to the customers of the two merging organizations. Successfully doing this was important to retaining customers in an extremely competitive financial services environment.

Potential to realize additional revenue streams through data services – One of the Vice President's charters within the company is to provide new technology solutions that either increase revenue or reduce cost. One option here is to create services that enable the business to monetize its data and applications and tap into previously unavailable revenue streams. An example is repackaging a business service in order to offer it through additional distribution channels (e.g., taking a billing system currently accessed through a local browser and making it available through an enterprise portal or as an app on a mobile device). The ability to build a data services layer using the Cisco Data Virtualization Suite now provides the company with an infrastructure that can support the exploration of these new marketing and revenue opportunities.

Cost-effective, consolidated infrastructure for data delivery – A key benefit is the fact that data virtualization consolidates the software licensing and physical infrastructure that would otherwise be associated with two distinct technology categories. One is technology for building data integration solutions, including data caching, while minimizing the need for an ETL-style infrastructure. The other is technology for building a services layer and APIs around data without requiring separate application servers and web servers. Thus, data virtualization offers a consolidated infrastructure that supports both of the company's strategic use cases and enables it to avoid the additional cost of dedicated infrastructure for ETL and/or web services. The Vice President added another important point here. "Once you implement a data virtualization solution, its cost-effectiveness increases every time you add more 'tenants' to it. Data virtualization becomes cheaper as you spread the cost across additional data sources and data consumers."

RETURN ON INVESTMENT. The Vice President could not share specifics regarding the ROI experienced with the data virtualization implementation, but it is clear from the benefits described above that data virtualization has significantly increased the company's business agility and IT flexibility.

Summary of Critical Success Factors

The firm's visionary information architecture, with sources decoupled from consumers to provide new levels of flexibility and agility, was the foundation for success. The benefits of this architecture led to the success of initial data virtualization implementations and the ability to subsequently extend the architecture across firms to support the merger challenge.

Several technical strengths of the Cisco Data Virtualization Suite were also critical factors that contributed to the company's success with data virtualization. These include:

- The ability to easily create virtual views across disparate data sources using Cisco's integrated development environment (IDE) facilitated rapid response to new information requests. This turned business users into data virtualization advocates.

- Cisco's ability to cache materialized views offered significant benefits, such as providing 24x7 access to data, minimizing the performance affect on underlying systems and the ability to update the cache on a scheduled basis. These effectively countered objections from ETL teams designed to justify the data-consolidation status quo.

- Cisco's ability to expose business functionality as web services eliminated the need to write special Java or web services wrapper code. This enabled the delivery of a wide variety of data sources – including legacy, external, merged and more – to the business in a consistent manner and further encouraged business adoption of data virtualization.

Future Directions

As mentioned above, current efforts are focused on increasing the acceptance of data virtualization as a viable solution for appropriate use cases throughout the now larger firm. The goal is to increase the scope and cost-effectiveness of the investment in the existing data virtualization layer by adding new data sources and data consumers.

The Vice President also indicated that data virtualization is becoming increasingly important in enabling analytics on very large data sets. Two potential benefits of using data virtualization here are the ability to incorporate a grid-type massively parallel processing (MPP) analytical data environment as a data source and the fact that Cisco can provide consuming applications with a SQL interface (rather than MapReduce) to access Hadoop data sources.

Global 50 Energy Company

Organization Background

This international oil and gas firm operates across six continents and its products and services are available in more than 100 countries. As one of the largest oil and gas producers, the company helps the world meet its growing need for heat, light and mobility, and strives to do so by producing energy that is affordable, secure and environmentally friendly.

For this case study, we interviewed the Head of Information Architecture, Upstream. (Upstream denotes the exploration and production component of the business). Information Architecture is the IT group within exploration and production responsible for developing an overarching architecture that integrates an information environment comprised of thousands of applications and users. The goal is to identify all of the business data involved and the tools, techniques and architecture necessary to manage that data, e.g., how to integrate, move, match, represent, report on and analyze data; improve data quality, etc. This IT executive's team includes data modelers and architects focused on both the logical side and the technical side. Functional scope encompasses master data management, data mapping, search, data quality, business intelligence, data storage, data integration, document and records management, portals and collaboration, data modeling, reference data architecture and metadata and taxonomies.

The Business Problem

An international oil and gas company operates within a very complicated business model. Exploration activities, for example, can involve tackling some of the most complex engineering problems in the world. An example is drilling an oil well through 5,000 feet of water and then another 35,000 feet below the ocean floor. In addition, the company's business processes are very diverse. As the IT executive described it, "We are not like a manufacturing organization where 80% of information needs may be met by an ERP application. We have some ERP, but there are a total of about 3,000 applications in our business portfolio. While our focus is global, there are local variants to everything we do. That plus a buy-versus-build approach have resulted in a very large application portfolio."

 The organization wanted to establish a single, logical source for all data access with the ability to return the data to the user in a view that conformed to a common business model regardless of where the data originated.

The key business problem was how to provide access to information in a complete and consistent fashion when data is stored and managed in multiple systems and locations. The organization wanted to establish a single, logical source for all data access with the ability to return the data to the user in a view that conformed to a common business model regardless of where the data originated.

The Data Virtualization Solution

The company's solution is a data virtualization layer implemented with the Cisco Data Virtualization Suite that provides a virtual data warehouse (DW) and virtual data marts to support the analysis,

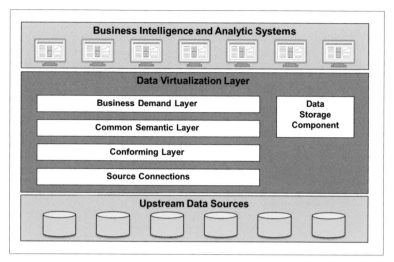

Figure 1. Global 50 Energy Company's Data Virtualization Architecture
Source: Global 50 Energy Company

reporting and decision-making needs of business intelligence and analytic systems. Figure 1 shows the logical architecture of the solution.

Data sources – Over 40 data sources are currently connected to the data virtualization layer and this number will grow over time. These sources represent a wide range of business functions from production and well data to people and finance data.

Data virtualization layer – This provides a single point of access and a single logical set of code to connect consuming applications to data sources. The current version of the data virtualization layer covers almost 600 common entities and over 7,200 attributes and the data model is increasing in complexity as new data sources are added.

The company has created a layered architecture within the data virtualization environment.

- Business demand layer – Publishes the views and web/data services that applications use to access the data.

- Common semantic layer – Provides a repository for views that represent the 600 shared entities of a common data model.

- Conforming layer — Transforms the data returned by the source connection(s) into views that conform to the common data model.

- Source connections — Access the data sources and make the source data available to the conforming layer.

- Data storage component — Enhances the performance of the data virtualization layer by staging data for faster retrieval using an IBM Netezza DW appliance. We describe this data storage component in more detail below.

Whenever data passes from the source through the data virtualization layer, it is conformed to a common semantic (metadata) model. "This is a critical component of our use case," said the IT executive. "We standardize not only on the physical infrastructure with data virtualization as a single source for data access, but also on ensuring that all data on the back end conforms to a certain semantic structure through the views regardless of where the data comes from. The application does not go directly to the system of record [the data source] but rather to the record of reference, which is the data virtualization layer. This is a powerful component of our architecture."

Consuming applications — Over 20 business intelligence and analytical applications currently use the data virtualization layer and the company plans to add 12 more by the end of 2011. Usage spans a wide range of application scenarios, from one user or a small group of users executing queries periodically to a dashboard that has thousands of users running multiple queries daily. Query volume has been as high as 28,000 per day and is expected to grow to 50,000 per day by the end of 2011.

COMMON CANONICAL DATA MODEL. The company has conformed the diverse upstream data into nearly 600 common canonical entities. The team logically modeled these entities using Embarcadero Studio. Next, the team built the layered set of views that bound them to the data sources using the Cisco Data Virtualization Suite.

INTELLIGENT DATA STORAGE COMPONENT. As described above, the company has extended the performance and scalability of the data virtualization layer with a data storage component implemented with an IBM Netezza DW appliance. According to the IT executive, "The data storage component is what really makes data virtualization work for us from a scalability and performance perspective. Netezza is essentially a massive engine underneath Cisco that makes it run faster."

Netezza can be configured to fire queries at any layer within the data virtualization environment to retrieve and store data that can then become an alternative source for queries. This is called "staging" the data.

Here is an example of how the data storage component augments Cisco. After connecting a data source to the data virtualization layer and creating a virtual view of the data, the administrator chooses to stage the data for the view and specifies the frequency for refreshing the data. Netezza then automatically archives the source data into a raw data table on Netezza. There is no need for the administrator to physically create tables or worry about structural changes on Netezza. This is all done automatically in the data storage device based on the view definition. It is common to stage source data in Netezza without any transformation at the source. Netezza then handles data transformation through queries to/from Cisco. Netezza fires a query at the demand layer and Cisco fires a fetch query back to Netezza to generate the transformed results. Netezza transforms the data and moves it from a raw table to a conformed aggregated table.

This transformed data is then used as an alternative source for queries. When a query hits the virtual view, the data virtualization layer knows if the data has been staged in the data storage component. If so, Cisco simply passes the query to Netezza, which has over 200 processors available to crunch the data, instead of the original data source.

Netezza can run queries through Cisco every minute, hour, day, week or month, or on a scheduled basis, to refresh the staged data. The staging frequency depends on the desired currency of the data. (As a note, if a query requires zero latency/near-real time data, the query is passed directly to the source data instead of Netezza.) Netezza is a more sophisticated alternative to Cisco's caching capability because Netezza knows how to handle schedules and dependencies and provides much higher performance and scalability.

According to the IT executive, "This whole process is so fast there is no need to create any data marts. The partnership between Cisco and Netezza is super powerful. The possibilities are endless and we are in the process of perfecting our solution. Version 1 of this capability has been running well for more that 16 months and is far more scalable for 28,000 queries per day than Cisco alone." He emphasized that both Cisco and Netezza are key to making this approach work well.

ADDITIONAL USE CASES. Beyond supporting business intelligence and analytics, the company uses the power of the data virtualization layer for connectivity and optimization to enhance agility in two additional areas:

Data source migration – If the data source is already connected to the data virtualization layer, the organization can swap out the back end, make changes to the common model views within the data virtualization layer, and then re-plug the conforming layer into the new source system without affecting the downstream applications in any way.

Application migration – Data virtualization enables the organization to avoid any disruption in service by directing queries from the new application to either the historical legacy system or the new system as appropriate. The application doesn't care where the data comes from and the old system can be retired at some point.

The Implementation Process

The data virtualization effort was driven by the Information Architecture group. The company set up what it calls a capability development activity project to instantiate the data virtualization architecture and capability as a service. The project included designing the architecture and processes, purchasing the hardware and software, writing the coding standards, etc.

As part of the project, Information Architecture spent significant time engaging the support of two groups. One was the company's central development shop because the project involved bringing responsibility for a new and different technology for data integration under the development umbrella. The other group was operations, staff responsible for running the applications and servers and handling support calls. The IT executive commented that, "As they look back now, from an operations perspective, the system runs much more smoothly now. Netezza runs lights out; it literally takes a DBA 10 minutes a day to support it, even with over 15TB of data and 600 tables. There are also hardly any support calls anymore."

> *Think outside the box. Don't believe what the old school data warehouse folks would tell you, namely, that a virtual data warehouse will never work. There is a huge shift in the market today.*

The data virtualization solution has been up and running since August, 2009, and has only been down once for three hours.

ADVICE. In terms of advice, the IT executive offered the following. "Think outside the box. Don't believe what the old school data warehouse folks would tell you, namely, that a virtual data warehouse will never work. There is a huge shift in the market today. It is not just the innovators such as Cisco that are focused on data virtualization. A number of large software companies are also getting on board with data virtualization. This is a direction enterprises should be heading."

Further, he counseled that it was a good idea to centralize initial design, development and deployment responsibility into a focused data virtualization team. "This allowed us to advance quickly and take on bigger concepts such as the common data canonicals and intelligent storage component and thus deliver a more powerful and complete data virtualization solution."

Summary of Benefits and Return on Investment

BENEFITS. The IT executive summarized in three categories the significant benefits the organization has achieved with data virtualization. Each of these categories is a major driver for the data virtualization effort.

Reduced risk through better decisions – By making information more easily available through faster delivery of integrated, high-quality information at less cost, data virtualization dramatically improves the organization's overall decision-making process. "We get better decisions that significantly reduce the risk of operations, and this is a major factor in everything we do," he explained.

Enhanced ability to take advantage of business opportunities to increase revenue and reduce costs – The integration of key data across the organization makes it much easier to identify and act on opportunities to increase revenue and/or reduce costs. An example is having the right data available to make good decisions about where to drill new wells, each of which has huge revenue potential. On the cost side, developing a major oil and gas platform is a large, multibillion dollar engineering project. Effective, integrated reporting on project progress enables the company to better control and reduce overall costs.

Another example of reducing cost comes from procurement. If the same supplier has different names within multiple disparate systems, it is difficult to view the company's relationship with the supplier in aggregate. An integrated reporting system based on a common data model enables the company to pull all of that information together on a global basis to identify everywhere it uses a particular supplier and the total amount it spends on that supplier. This increases leverage and facilitates better control over supplier costs.

Improved efficiency and resource allocation to enhance agility and competitiveness – "The ability to do things faster and smarter gives us a competitive edge," said the IT executive. As one example, automating the delivery of information to eliminate manual reports saves a lot of time and effort that can be reallocated to more valuable activities. "Industry metrics document that engineers spend an average of 40 percent of their time just gathering data. If we can reduce that, we allocate our resources more intelligently and work far more efficiently."

Improved IT performance to reduce IT costs – From an IT performance point of view, data virtualization also delivered a number of specific benefits including:

- Reduced time to solution for meeting new business requirements.

- Much lower total cost of ownership than any previous approach, including the minimal support staff required.

- More readily accessible environment supports reporting and analytic tool flexibility.

- Simpler, more agile architecture (e.g., the need to connect to each data source only once).

- High performance through the combination of Cisco and Netezza.

RETURN ON INVESTMENT. The organization has reduced overall development costs by 40% through code reuse on both the front end (data and web services, sharing of common model views of data sources) and the back end (connectivity to the data sources only has to be defined once). For applications that don't already fit into the common model, there may be a slight bit of overhead required to implement the application initially.

From a data integration maintenance and support perspective, data virtualization requires far fewer resources. For example, data integration support for one major application required two full-time equivalents (FTEs) prior to data virtualization. Migrating the application added a mere two hours of support per week in the data virtualization environment, a net savings of at least 78 hours every week. "So we just added those two hours per week to the work load and it didn't even cause a blip," stated the IT executive. Support for the entire data virtualization environment currently requires only 1.5 FTEs.

Summary of Critical Success Factors

Several aspects of the data virtualization solution were critical to the company's overall success.

Achieving the required performance through the data storage component – A major success factor was the ability to combine the agility of Cisco with the power of Netezza in an easy-to-use manner. "We were able to resolve the performance issue of data virtualization when very large volumes of data are involved. Our solution would not have worked otherwise," stated the IT executive.

Accepting the concept of data virtualization – Adopting data virtualization requires separating how data is accessed from how the organization uses the data. The IT executive put it this way: "It separates the 'what' of understanding the information itself from how we manage and store the data."

Agreeing on and implementing a common data model – The common canonical data model ensures the data provided through the data virtualization layer is consistent and high quality. This gives the business users confidence and makes the IT staff far more agile and productive.

Centralizing the development team – The development team was comprised of 30 people, 10 onshore and 20 offshore. This balance helped speed development and reduce time to solution with a round-the-clock development effort. Another aspect of this was separating the responsibilities for architecture and development within the team.

Management support – High-level executive buy-in was very important.

Continued marketing of the concept of data virtualization – The marketing effort has proven to be a key to broadening support for data virtualization throughout the organization. "In fact, sometimes people say, 'Why didn't we hear about this sooner?'"

Internally branding the solution – "While branding has advantages and disadvantages depending on your track record, branding worked for us because our implementation has been successful and given the brand a 'good name.'"

Future Directions

The IT executive cited two areas where the company plans to build on its data virtualization solution going forward.

• Leverage data virtualization for content mapping, hierarchy management and master reference data.

• Leverage data virtualization for data archiving and data migration.

Global 100 Financial Services Firm

Organization Background

This data virtualization user is a leading global provider of financial solutions for corporate, institutional and private clients. The company has two divisions. The investment banking arm comprises the bank's capital markets business. The private asset management arm includes the investment management business for both private and institutional clients plus the company's retail banking business. This case study focused on the firm's investment banking division.

For this case study, we interviewed Marc Breissinger, Executive Vice President at Cisco. He has been with Cisco for more than three years and is one of the primary architects of the investment bank's Data Vault project, the subject of this case study. Breissinger works closely with the bank's Enterprise Architecture team on this project.

The Business Problem

In today's competitive environment, investment banks face significant challenges that revolve around the increasing volume, velocity, complexity and timeliness of their business processes and data.

- Volume – Trading volume and the amount of detailed data that must be managed is constantly growing.

- Velocity – The rate at which the data changes is increasing dramatically.

- Complexity and integrity – Regulatory requirements imposed on investment banks are not only changing fast but growing in complexity, particularly following the financial collapse of 2008. More data has to be kept for longer periods of time, the auditing process is complicated and costly and the penalties for noncompliance are stiff. Data integrity and accuracy are critical to meeting these requirements.

- Timeliness – Both internal and external customers increasingly want real-time access to data on demand. The difficulty inherent in delivering real-time data is exacerbated by all of the factors above.

The 2008 financial crisis was a strategic watershed for investment banks. According to Breissinger, the company's investment bank division "took a step back and surveyed the marketplace to assess the best option going forward. One was to simply pull back into its shell, cut costs and ride out the financial storm. Instead, the bank decided to invest in a new, strategic IT infrastructure to better address these complex data requirements in the future. The company viewed this as an opportunity to leapfrog the competition by changing the way the investment bank's IT infrastructure worked."

Breissinger offered some perspective on this. "Investment banks tend to think of their product as software. Typically, this starts as an idea to execute something very quickly – faster than the competition – to get a leg up on the market. So when someone has an idea with the potential to do this, such as a new trading strategy, an investment bank typically creates a vertical IT infrastructure to support the implementation and ongoing support of the idea. Thus, over time, most investment banks end up with multiple, vertical silos and each silo has its own mini IT organization."

As an investment bank grows, however, this process becomes unsustainable for several reasons. One is the duplication and inefficient use of infrastructure. The operating cost of maintaining all these IT organizations is "staggering," said Breissinger. The vertical silos also create problems in data consistency and incur significant data reconciliation costs. "When a new application team needs data from an existing team, the answer usually is an extract of data copied to the new silo and transformed along the way because the source system, from a performance perspective, cannot support two masters – both updates and queries. And these copies of the data proliferate. One application creator makes copies for four or five other applications, each of those propagate the data again multiple times, and so on."

The firm estimated that there could be up to 30 copies of every piece of source data floating around in one form or another. This problem of "uncontrolled replication" had serious ramifications for the bank. Timeliness of data was an issue because every copy operation introduced latency. Data decay was an issue because each application transformed the data a little differently. "When you are working with data that is eight transforms down the chain, you have a huge problem verifying the lineage of the data on a report to a regulator. People are not only making decisions based on old data of unknown quality, but also making decisions based on data they cannot prove is accurate to a regulator."

> *"Two information architecture strategies were critical ... First, drive all data consumers to use one common, logical version of the source data. Second, eliminate uncontrolled replication of source data. To implement these strategies, the investment bank leveraged data virtualization.*

To address these and other issues, the bank developed a four-year program called "SI 2012." SI stands for "simply instantaneous" and the target date for implementation was 2012. The objective of SI 2012 is to provide instant access to information wherever and whenever needed. This requires having the infrastructure in place by the end of 2012 to enable any data consumer to access any critical bank data in real time.

Two information architecture strategies were critical to the success of SI 2012. First, drive all data consumers to use one common, logical version of the source data. Second, eliminate uncontrolled replication of source data. To implement these strategies, the investment bank leveraged data virtualization. The goal was to "move the process to the data rather than move the data to the process," explained Breissinger.

The Data Virtualization Solution

The investment bank's data virtualization solution is the Data Vault, an operational data store (ODS) designed to provide data consumers with real-time access to all of the division's shared data. The Data Vault represents the "golden source copy" of data, or the single version of the truth, for this data across the entire investment bank. A data virtualization layer implemented with the Cisco Data Virtualization Suite sits between the data consumers and the ODS to create a single, consistent virtual view of the data across all of the data repositories and components within the Data Vault.

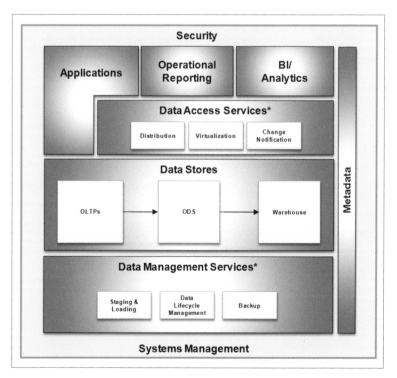

Figure 1. Global 100 Financial Services Firm Data Vault Logical Architecture
Source: Global 100 Financial Services Firm

The overall logical architecture of the Data Vault is shown in Figure 1. The Data Vault is designed to be distributed, highly redundant and consistently available. Here are the primary components of the architecture.

Data sources – Data sources include any systems that store and manage shared data, such as trading and asset management OLTP databases. Data from each source is replicated and transformed only once, and only into the ODS within the Data Vault. This is done automatically using one of the bank's primary replication mechanisms (change data capture, messaging and ETL) whenever a transaction occurs in the source.

Operational data store (ODS) – The ODS is the Data Vault repository for source data. While source data can be copied only once, the architecture does permit data to be replicated, partitioned and/or moved *within* the Data Vault to provide the performance, scalability and availability necessary to meet service level agreements (SLAs). For example, to ensure availability, there are three copies of the ODS, each in a different geographic location. These copies are referred to as "central functions." At any point in time, one central function is the active copy, one is the disaster recovery copy and the third is the maintenance copy. So even during maintenance cycles, there are always two current copies available. These roles are rotated on a quarterly maintenance schedule, described below.

Edge servers are satellite Data Vault locations that represent distributed components of the ODS. Subsets of the data and services from the central function are copied to and cached in each edge server to provide better performance for locally attached data consumers. The combination of a central function and its edge servers is referred to as a "stack."

Breissinger described the relationship between the central function and the edge servers as follows: "Within a stack, we replicate the data once from the source into the central function. For any subset of data configured for replication to the edges, changes are propagated to the edge where they are then applied to the cached subset of data. The ability to send only changes minimizes the data flowing over the network and keeps the data up to date in near-real time." This all happens internally within the Data Vault and is transparent to the user. The overall goal is controlled replication within the Data Vault with a finite number of targets that are updated in near-real time.

Data warehouse – Another component of the Data Vault is the data warehouse, which contains summarized data derived from the ODS.

Data virtualization layer – The data virtualization layer provides the single virtual view of the underlying Data Vault repositories – the transactional stores, the ODS and the data warehouse – for data consumers.

The data virtualization layer has three major functions: (1) manage the views and services through which data consumers access the Data Vault, (2) cache and distribute ODS data in the form of materialized views to improve performance (caching is done both centrally and in the edge servers), and (3) manage the data replication and change notification processes.

Consuming applications – Operational reporting, BI/analytics and most other business applications will be affected because they must go through the Data Vault (directly or via edge servers) to access shared bank data. The goal is to provide a set of standardized services with a common interface for accessing the Data Vault.

Other components – The Data Vault also encompasses common security and systems management components plus shared metadata and data management services (e.g., staging and loading data, backup, etc.).

DATA UPDATE PROCESS. The Data Vault is architected to update the ODS with changes to source data in near-real time. These changes to the ODS are then captured so that relevant changes can be sent to the change notification component in the central function data virtualization layer. Then, depending on the views and cache configurations and change notification rules, the central function updates the associated views and caches as well as notifies the consuming application that data has changed.

ODS MAINTENANCE PROCESS. Another interesting aspect of the Data Vault is the quarterly maintenance process for the three central ODSs. This process essentially implements a new "release" of the Data Vault, including hardware and software changes and upgrades and changes to source data structures and views. The current maintenance copy of the central function is taken offline and upgraded to the new release. Then IT runs tests for about two weeks to verify that everything works as expected. At this point, the central function roles are switched. The old maintenance copy is now one release ahead and becomes the active copy. The old active copy becomes the disaster recovery copy, and the old disaster recovery copy is now the maintenance copy. This maintenance copy is then upgraded and becomes the disaster recovery copy, and the remaining copy is upgraded and becomes the maintenance copy going forward.

CURRENT STATUS. The firm is still in the process of implementing the Data Vault. Currently there are three edge servers in the same locations as the three central functions. Over time, additional edge servers will be implemented in other distributed locations to optimize query performance. Each ODS consists of one physical database instance but the architecture allows for multiple instances in the future. Data virtualization would then provide a single, abstracted view of the data across all instances in each ODS.

The Implementation Process

The concept and architecture of the Data Vault was defined as part of the SI 2012 program that the firm launched in 2008.

Because data virtualization was already in place supporting other use cases, it was evaluated as part of this larger initiative. The proof of concept that formally evaluated how data virtualization would enable the Data Vault occurred during 2009. As a result of this evaluation, a number of enhancements to the Cisco Data Virtualization Suite were identified to support the distributed, global architecture and data volumes. These enhancements, including incremental caching, distribution and monitoring of caches across multiple Cisco servers, change data capture integration, new performance optimizations and several other new features were developed during 2010.

By early 2011, the pilot Data Vault implementation that used these new capabilities was in place. According to Breissinger, the Data Vault is only one component of SI 2012 and does not, in itself, achieve the overall goal of the program. While the Data Vault addresses data replication and availability issues, significant process reengineering, information modeling and organizational restructuring is also required to fully implement the SI 2012 solution.

> *It is OK to start small with point implementations. This will accelerate business benefits and help fund the larger, more ambitious data virtualization deployments that will surely follow.*

ADVICE. Because the Data Vault is a globally distributed solution that also integrates replication and messaging capabilities, it is an ambitious, push-the-envelope enterprise information architecture. Further, the extensive SI 2012 program required a multiyear deployment. Data virtualization deployments of this scope are not the typical starting point. "It is OK to start small with point implementations," said Breissinger. "This will accelerate business benefits and help fund the larger, more ambitious data virtualization deployments that will surely follow."

Breissinger also added, "In an enterprise-wide effort of this scope, the majority of challenges that must be overcome for success will be organizational and not technical. Moving from a very vertical, siloed organization model for data management to a horizontal, shared model requires a realignment of incentives and an aggressive leadership approach."

Summary of Benefits and Return on Investment

BENEFITS. Breissinger described several expected benefits of the Data Vault data virtualization solution.

More effective business decisions and market insights – Easy, real-time access to a high-quality, comprehensive view of current investment bank data enables data consumers to focus their efforts on effectively analyzing the data in order to gain insights and make better business decisions, rather than on non-value-add activities such as finding and reconciling the data.

Flexible, efficient IT infrastructure – Because the data virtualization layer abstracts both the data in the underlying repositories and the physical implementation of Data Vault components, IT can make changes and enhancements without affecting the view seen by data consumers. In addition, creating an infrastructure that allows only a single copy of each source data, and one owner of that data, eliminates all of the problems and inefficiencies inherent in an environment of uncontrolled data replication. This single view of investment bank data and the ability to create standardized services to access the data also reduces the cost and time required to develop new applications and reporting systems.

More cost-effective and efficient use of application development resources – Prior to implementation of the Data Vault, each primary (source) application team had to spend up to 70% of its resources on providing data extracts for other teams rather than on improving its own applications. Now, instead of creating multiple extracts, the team can set up a single, continuous data feed to the Data Vault and redirect development resources to activities that generate more value. The Data Vault also improves productivity by eliminating the need to spend time reconciling inconsistent views of data.

RETURN ON INVESTMENT. The firm's SI 2012 initiative has an overall investment of €100 million. Data virtualization and the Data Vault are the critical foundation for achieving a significant return on this investment. To date, returns have focused on the cost savings in development resources and infrastructure. Anticipated business benefits, including support for business growth and improved decision making, will soon be realized.

Summary of Critical Success Factors

The most important factor in the success of this initiative has been close cooperation between the investment bank and the data virtualization solution provider, Cisco. This cooperation resulted in a partnership to design and fund the development of several important product enhancements that were needed to meet specific Data Vault and SI 2012 requirements.

A second critical success factor has been the quality of the people involved. To support this partnership, both sides enlisted their top architects and technologists. This "A-team" approach enabled conceptualization and delivery of this challenging, first-time-ever architecture.

Finally, flexibility has also been a critical success factor. From a data virtualization product point of view, Cisco had to be flexible to extend its offering to meet the firm's advanced requirements. The firm had to be flexible in defining a schedule that would work for both parties. And looking ahead, the architecture of the Data Vault will need to continue to be flexible to meet new business requirements and take advantage of new technology advances.

Future Directions

Going forward, the first priority remains successful, global deployment of the Data Vault initiative in support of the SI 2012 program. This will include work on multiple fronts including data modeling and other data governance activities, a phased transition of data sources and consumers in support of the Data Vault and on-going refinement and expansion of the overall architecture.

Northern Trust

Organization Background

Northern Trust is a leading provider of innovative fiduciary, investment management, private banking, wealth management and worldwide trust and custody services. Clients include corporations, institutions, families and individuals. The company has remained independent since its founding in 1889 and is positioned within selected individual and corporate market niches as a well-respected provider of trust and investment services with an emphasis on strong relationships.

Northern Trust is a financial holding company based in Chicago, Illinois. It serves clients in more than 40 countries and has over 13,000 employees in offices in 18 U.S. states and 16 international locations in North America, Europe, the Middle East and the Asia-Pacific region. At the end of 2010, the company had assets under custody of $4.1 trillion, assets under management of $643.6 billion and banking assets of $83.8 billion. Annual revenues in 2010 were almost $3.7 billion.

For this case study, we interviewed Leonard J. Hardy, Senior Vice President, Operations and Technology. Hardy is a member of Northern Trust's Rapid Solutions group, which provides a solution architecture and consulting facility to other Northern Trust application development groups. If one of these has a difficult technology or business problem, the Rapid Solutions group has ten diverse sets of people that can help solve the problem. Hardy also manages the company's Integration Competency Center (ICC), which helps other application areas figure out the best way to integrate data they do not own. The ICC supports several tools for this, including the Cisco Data Virtualization Suite, ETL technologies and data integration best practices. Hardy participated in this case study project in his ICC role.

The Business Problem

An important line of business for Northern Trust is providing outsourced investment management operations for corporate customers, typically investment management firms and banks. These institutions outsource components of their internal investment management office functions to Northern Trust. Northern Trust then provides those functions for the institution's investment clients who are typically high-net-worth individuals.

The benefits for these financial institutions can be significant. By contracting these functions out to Northern Trust, the institution does not have to invest in the resources, assets and skills necessary to provide the functions internally, instead leveraging the capabilities of Northern Trust. In return, Northern Trust provides guaranteed levels of quality, service, resilience and value-to-cost criteria and management.

The institutional customer continues to provide front-office investment management functions for its clients, but outsources all or some of the middle-office and back-office functions (see Figure 1). In the case of full outsourcing, as soon as a trade is executed for a client, the transaction is sent to Northern Trust which handles all subsequent processing, accounting, reconciliation and reporting.

The Northern Trust business unit responsible for investment management outsourcing is Corporate and Institutional Services (C&IS), the corporate arm of the company. The line of business within C&IS is called Investment Operations Outsourcing (IOO).

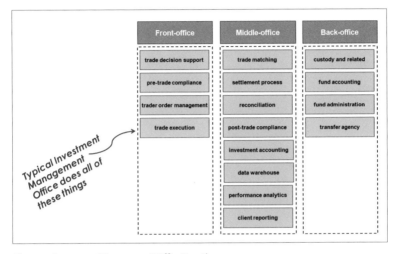

Figure 1. Investment Management Office Functions
Source: Northern Trust

> *If we could get customers on board more quickly, we would realize several key benefits. First, revenue would increase. Second, we could go down the pipeline faster ... and take advantage of huge new business opportunities in the marketplace.*

The business problem for IOO was the fact that it had a large number of new institutional customers in the outsourcing pipeline and simply could not implement them fast enough. As Hardy stated, "If we could get customers on board more quickly, we would realize several key benefits. First, revenue would increase. Second, we could go down the pipeline faster. We had customers already signed up for outsourcing who were waiting to be converted and we had to find a way to do that faster. That, in turn, would allow us to open up the pipeline and take advantage of huge new business opportunities in the marketplace."

One of the biggest obstacles was the time it took to set up the end-client reporting function for each customer so Northern Trust could communicate data, such as performance and valuation reports, back to the institution and to its end clients. Existing IOO client reporting capabilities had evolved within a traditional systems infrastructure and were both inefficient and inflexible. For example, value-added client data was stored in multiple separate databases, including legacy, mainframe systems. In addition, some of the business logic to interpret the data structures and create reports was embedded in stored procedure code. The only way to access the data and interpret it correctly was to go through the stored procedures. There was no data warehouse or data mart that consolidated all this data and business logic for reporting purposes.

As a result, according to Hardy, "the old way of reporting required technology intervention for every single customer implementation. As you can imagine, each institution has its own formats, labels, graphics and fonts. A technology expert proficient in our development tools had to get involved and make changes to our reporting process for every customer. The time to market on that was very long. We needed the ability to bring on new customers without having to make technology changes for each one."

To complicate matters further, Northern Trust had developed a three-year client reporting strategy to meet future data and reporting needs across all of its corporate and business customers, including IOO. IOO could not wait that long to solve its own critical need for a new client reporting architecture, yet any solution IOO implemented now would have to support the long-term strategy.

Finally, the company had already purchased a client reporting tool. This tool assumed the existence of a data warehouse to consolidate and produce all of the client data. The tool did not have the ability to access multiple data stores and merge them together to produce a unified view.

One solution for IOO was to physically integrate the data using ETL and data warehousing technology. But that would require pulling all of the business logic out of the stored procedures and figuring out how to get all of the data into one database and have the structure make sense, a massive project that could take up to a year and a half.

Instead, IOO took a more innovative approach to solving the business problem and decided to virtually integrate the data using the Cisco Data Virtualization Suite as a virtual data warehouse. Placing Cisco in the middle between the client reporting tool and the back-end data sources would enable IOO to quickly abstract the data into the data virtualization layer, reuse the existing business logic to access the data and generate all of the data needed for end client reporting.

As a note: This project addressed how to provide outsourcing customers with the functionality represented by the bottom three boxes on the middle-office tier as shown in Figure 1: data warehouse, performance analytics and client reporting.

The Data Virtualization Solution

Northern Trust's IOO data virtualization solution is a virtual data warehouse implemented with the Cisco Data Virtualization Suite

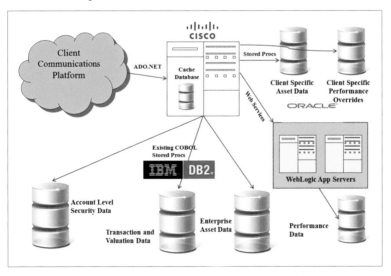

Figure 2. Northern Trust's Investment Operations Outsourcing (IOO) Architecture
Source: Northern Trust

combined with a new client reporting front end. Here are the major architecture components (see Figure 2).

Data sources – There are several physical data stores in the IOO environment.

- Transaction data includes all of the detailed transactions for each end-client account: buys, sells, interest income, dividends, etc.

- Valuation data includes, for each end-client account, how much it is worth today, how much it was worth yesterday, etc. based on the transaction data.

- Enterprise asset data is a reference file that contains data about each asset in the client accounts, such as ticker symbol and investment category (equity, bond, money market, etc.). This data comes from a number of different third-party providers.

- Client-specific asset data customizes the enterprise asset data based on institutional customer preferences. Some customers do not like the name, classification, etc. that comes from the enterprise asset database and will provide substitute data where appropriate.

- Client-specific performance overrides also customizes data based on institutional customer preferences. Some customers prefer to use different or special indexes against which to compare client account performance.

- Performance data includes very complicated investment performance calculations that permit performance comparisons between accounts, such as one account against accounts of similar value or with similar investment strategies. Each institution determines how this information is presented to its clients.

Transaction, valuation and enterprise asset data is stored in IBM DB2 and accessed via COBOL stored procedures. Client-specific asset and performance overrides data is stored in Oracle and accessed via Oracle stored procedures. Performance data is also in Oracle and accessed via web services that call C code.

Data virtualization layer – The virtual data warehouse built using the Cisco Data Virtualization Suite abstracts the data in the IOO data stores and creates a unified, virtual view that makes them appear as a single physical data store. For performance reasons, Cisco caches data that doesn't change frequently (such as enterprise asset data) to avoid accessing the source data where possible. When the reporting platform issues a query to retrieve data from one or more data stores, Cisco queries the data by calling the appropriate stored procedures and/or web services, combines the data (for example, associates all of the transactions that make up a valuation and the important performance data that applies to that valuation) and delivers it to the reporting platform.

Consuming applications – The primary consuming application is the client reporting suite. This is the toolset that IOO's internal business/operations partners now use to build, format and produce the reports for IOO's customers and their clients. In addition to reporting, it has built-in workflow capabilities to manage and control the overall client communication and reporting process. A simple example would be a workflow to generate a report, send it to someone who adds manual commentary about an investment and then send the annotated report to the person who approves the report as customer ready. The report is delivered through a web interface to the customer who then can review the report before releasing it to clients.

Another benefit of the reporting tool is that it supports a data dictionary and business views of the data that directly map to views within the data virtualization layer. These business views, or building blocks, conform the data into reusable components that the business analysts can use to create custom reports.

A key point here is that the solution has enabled IOO to transfer the time-consuming report customization work that previously had to be done for each new institutional customer by very high-priced and very busy technical resources (applications programmers) to IOO's operations partners (business analysts). As Hardy summarized it, "We have been successful in our goal of taking the technology out of the outsourcing equation. As long as we supply the right data through Cisco, the reporting tool handles all formatting, graphing and other functionality that we used to have to hard code within the application program. One reason we had such a big bottleneck in implementing outsourcing customers was the fact that we didn't have enough programmers in our application development group to handle the customization requirements. Business analysts in our operations group can now easily process the building blocks necessary to customize reports."

> *We have been successful in our goal of taking the technology out of the outsourcing equation. As long as we supply the right data through Cisco, the reporting tool handles all formatting, graphing and other functionality that we used to have to hard code within the application program.*

By combining the capabilities of data virtualization and the new reporting tool, IOO has been able to navigate through and abstract a very complex set of layers of translation to connect the raw IOO data structures to business metadata and finished customer reports.

The Implementation Process

In the case of IOO, the first decision was to implement a data virtualization solution rather than physically integrate the data using ETL and data warehousing technology, as described earlier. This decision was based on the fact that physical integration would take more time and development resources (e.g., elapsed time would be an estimated 18 months versus about 7 months) and the fact that Northern Trust had already successfully used Cisco for other applications. Going forward, the company will extend its use of data virtualization as it builds common data services and identifies additional applications that can take advantage of these services.

When creating the data virtualization solution, IOO tried to include any data that institutions might want to report to their end clients. The company also recognized that as new institutions came on board, it might be necessary to tweak the views. Once the data virtualization layer is comprehensive enough, "all of the customization work will happen at the reporting tool level," said Hardy.

IOO has completed four institutional customer implementations with over ten more in progress or planned.

ADVICE. Hardy offered some lessons learned based on his experience working with IOO.

Understand the data – According to Hardy, data virtualization is no different than any other development or physical data integration tool. "The design and architecture of the solution is where you need to spend time. The old carpenter's adage of 'measure twice and cut once' applies here. Don't just dive in and start coding views. Make sure you really understand the data. This is the key to success in an effort like this."

Educate and support the business – Hardy said it is also important to allocate time to consult with the business and make sure they understand the data. "In our case, they are the ones designing the reports but they don't necessarily have in-depth knowledge about the data in its raw form. So we need to help them understand the raw data and make sure that when the data gets to the end tool, people understand what they are looking at."

It is equally important to be prepared to provide support if the user has questions. "Suppose an end user is building a report and one of the numbers doesn't look right. Is the problem in the reporting platform (the data was manipulated incorrectly), Cisco (the data was joined incorrectly), the stored procedure or web service (the business logic is incorrect) or the underlying base data? We had to set up a process to triage these problems, make educated guesses and get the right people involved at the right time."

Manage business expectations – Although data virtualization is a capability that can get the job done faster, the organization still has to go through the same steps to implement the solution: analysis, design and architecture. "In our case, we could say it saved us a year and a half, but it did take seven months. So it will take time no matter how you do it."

Summary of Benefits and Return on Investment

BENEFITS. Hardy described the major benefits Northern Trust has achieved with the new IOO client reporting architecture.

Faster time to market increases customer satisfaction and revenue – Data virtualization has dramatically reduced the time it takes to implement a new outsourcing customer. Moving customers through the pipeline faster improves overall customer satisfaction and gives IOO the capacity to bring in even more business and revenue.

Flexible infrastructure enhances agility and reduces cost – There are several aspects to this. One is the fact that the reporting tool, or any consuming application, has a single data access point regardless of the format, access method (SQL, web service, stored procedure, etc.) or data source. This significantly reduces application complexity and development and maintenance time and cost. The ability to create reusable data services that can be shared by all applications also reduces costs here.

In addition, Northern Trust is embarking on a three-year project to replace all of its underlying data stores with modern data warehouses and data marts. By putting data virtualization in the middle, thereby decoupling the backend data from the consuming applications, IOO will not have to change anything that has been defined in the reporting tool when it swaps out the back end data stores. It will only need to change the views to consume the new data sources. This flexibility will make it much easier and faster to migrate the reporting tool to new data sources.

RETURN ON INVESTMENT. The data virtualization solution has enabled Northern Trust to greatly reduce the time it took to solve the business problem with a 50% reduction in time to market and a 200% ROI. According to Hardy, "We did it all within seven months with 2.1 full-time equivalents in terms of staffing. Without Cisco, we estimated this would have taken a year and a half."

Summary of Critical Success Factors

There were several aspects of the solution and the implementation process that were critical to Northern Trust's success in this effort.

Start with a focused project – Hardy stated that a focused initial effort is key to success. "Don't start with a project to make all enterprise data available via a service."

Centralize support for data virtualization and development – Giving ICC responsibility for the data virtualization technology provided economies of scale and enabled the company to accelerate up the best practices learning curve in effectively implementing the technology in business solutions.

Take advantage of vendor professional services – Northern Trust also had a senior level Cisco consultant/architect on site for the first year to help with the more difficult design and optimization challenges, avoid pitfalls and resolve issues more quickly. According to Hardy, "Cisco's Professional Services was indispensable in getting ICC up to speed in data virtualization development and support early on. We have been able to phase out the need for that as we built up our own expertise."

Future Directions

As described above, as the three-year plan to migrate to new data warehouses and data marts progresses, IOO will adapt its data virtualization implementation so no changes will be required in the consuming applications. IOO is also enhancing and reusing the data virtualization layer for other applications that need access to valuations and performance data. For example, Northern Trust has other corporate customers who currently get a transmission of data each night or monthly in a computer-readable format. "We could use the existing data virtualization views to pull all the data together and then have the application format it and send it to customer. This would involve building a new, high-level view that calls the appropriate underlying views. This hierarchy of views enables us to use the same underlying virtual layer to service new requirements and customers."

NYSE Euronext

Organization Background

NYSE Euronext operates the world's leading and most liquid equities and derivatives exchanges. The company is comprised of six equity exchanges and eight derivatives exchanges located in the U.S. and Europe, including the New York Stock Exchange (NYSE), NYSE Arca, NYSE Liffe, Euronext and American Stock Exchange (AMEX).

Over 8,500 issues are listed on NYSE Euronext exchanges and cover an extensive and diverse set of products, such as stocks, bonds, exchange-traded funds (ETFs), exchange-traded notes (ETNs), options, open funds, warrants, commodity futures and other derivative products. The exchanges handle over four billion transactions per day with an average daily value of $153 billion.

The NYSE was first organized in 1792, incorporated in 1971 and has evolved into NYSE Euronext through a series of mergers and acquisitions.

For this case study, we interviewed Emile Werr, Vice President of Global Data Services. Global Data Services evaluates, designs, develops and implements services, technologies and architectures for the entire firm. Areas of responsibility include business intelligence; common data services; data architecture; data access and integration; data warehousing; ETL/ELT design, development and integration; and grid/cloud computing. The team's scope is managing post-trade data, that is, data immediately after a trade executes through to data warehousing, data delivery, reporting and analysis.

Werr is also the Chief Data Architect with overall responsibility for all of the firm's data and the systems to support that data. In these complementary roles, Werr is part of the internal IT group while providing support for NYSE Technologies, a separate NYSE Euronext business that markets additional services to external member firms, such as investment banks and brokerages, banks and broker-dealers. Werr helps in the selection of solutions to help NYSE Technologies generate business opportunities.

The Business Problem

A major challenge for NYSE Euronext is the sheer complexity of its business and operating environment. There are several contributing factors here.

- First, the organization has gone through many mergers and acquisitions in its evolution. This has resulted in significant complexity in terms of the ability to effectively integrate data across multiple enterprises and exchanges.

- Second, NYSE Euronext trades 14 different types of products, "from the pure vanilla equities market to complicated derivatives plus commodities and futures," said Werr. "These all have different data structures, increasing our integration challenge."

- Third, NYSE Euronext deals with massive data volumes, producing an aggregate of 2TB per day, on average, across all of its exchanges and markets. During peak periods, volume can exceed 5TB per day. "This adds all sorts of challenges around ensuring that we can ingest that much data on a daily basis, synthesize it, get it into a data warehouse and make it consumable by downstream applications for regulatory, research, capacity planning, marketing and many other purposes."

- A fourth requirement is the need to meet rigid service-level agreements (SLAs) with business units for delivering and retaining data. Thus, it is difficult work just to "keep the lights on."

From a strategic point of view, NYSE Euronext is also involved in redefining the purpose and scope of an exchange in an effort to transform the business into a broader services provider. "We are an exchange but at the same time we are getting into other businesses, such as technology and data services," explained Werr. One example is NYSE Technologies' Secure Financial Transaction Infrastructure, or SFTI (pronounced "safety"), a global securities trading network that gives member firms access to a wide range of markets and services through a single connection point. SFTI is a SaaS (software-as-a-service) solution designed to provide customers

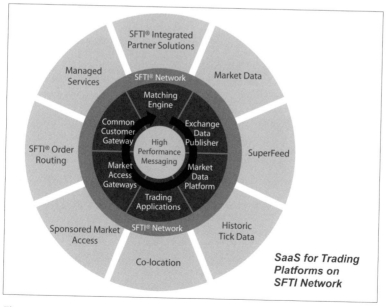

Figure 1. NYSE Technologies SFTI Trading Platform and Services
Source: NYSE Euronext

with professional services built around the high-performance, secure and reliable SFTI trading platform (see Figure 1). Services include co-location in the NYSE Euronext data centers for integrated trading activities and the delivery of tailored market data and historic tick data.

> *The need to remain innovative in a competitive environment while transforming the business exacerbates the inherent overall level of complexity. In addition, all of the information generated by these new business services has to funnel into the NYSE Euronext data delivery environment.*

"Our goal is to provide new products and services for existing customers and to find new customers for our products and services." Because customers have options for where they can trade securities, NYSE Euronext has to be innovative in order to attract trading to its exchanges, along with the associated fees and opportunities for other forms of revenue. The need to remain innovative in a competitive environment while transforming the business exacerbates the inherent overall level of complexity. In addition, all of the information generated by these new business services has to funnel into the NYSE Euronext data delivery environment. As a result, according to Werr, "Automation, standardization and performance have become critical to our success in meeting these objectives."

Werr's team established a number of initiatives in the areas of data integration and data delivery to address the business challenges. The first is to standardize access to data across the organization in terms of both architecture and toolsets. Key solution requirements are a consolidated data access and integration layer, and a set of tools that make it easy for business analysts, instead of technical staff, to maintain the logical data layer to take advantage of in-depth domain knowledge about business processes and data.

Standardized data access is then the core foundation for achieving follow-on initiatives such as integrating enterprise reference data and streamlining the enterprise reporting system.

The Data Virtualization Solution

The NYSE Euronext solution is an enterprise-wide data virtualization layer, built using the Cisco Data Virtualization Suite, that functions as a virtual data warehouse (DW) to provide read-only access to post-trade data for analysis and reporting. Werr has named the solution TORQCA (pronounced "torkah"), an acronym for the major data transactions that comprise the business: trades, orders, reports, quotes, cancels and administration (admin messages). The logical architecture for TORQCA is shown in Figure 2. Here is a summary of the components of the solution.

Data sources – These include both transactional systems and reference data systems. Examples of transaction data are quotes, orders, trades, acknowledgements and receipts. Reference data includes listings and member data, customer and products data in terms of the instruments traded, as well as data about corporate actions which are very important to the business (e.g., an announcement that affects a stock price, such as a management change or change in a dividend payment).

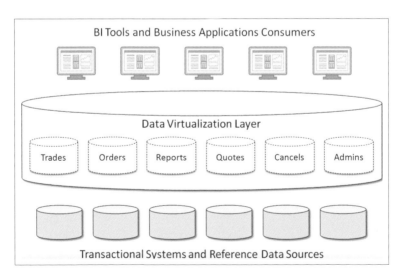

Figure 2. NYSE Euronext Reference Architecture for Data Virtualization
Source: NYSE Euronext

NYSE Euronext has multiple data centers located around the globe (each exchange has its own data center) and many different DWs and data marts. Given the volume of data and the strict SLAs for retaining data online (data needs to be quickly accessible for at least seven years to meet regulatory requirements), multiple physical DWs often exist to store different time slices or levels of granularity of the same data.

Performance is very important, not only because of data volumes but also because response time for analytics has to be fast. The company, therefore, is a strong adopter of massively parallel processing (MPP) technology.

Data virtualization layer – In the NYSE Euronext environment, the data virtualization layer has many functions and roles. It provides all of the following:

• Standardized data access for connectivity to all data sources.

• Virtual DW for federating data through logical views.

• Centralized repository for common metadata, application/ business logic and connectivity and data services.

Changes to any of these centralized components are made in the data virtualization layer, eliminating the need to redeploy code to applications. This makes it easier to embed business requirements directly into the development life cycle without going through multiple layers of translation at the application level. The data virtualization layer also makes the environment extensible and provides significant opportunities for reuse. These are all important drivers for faster systems development going forward.

When the data virtualization layer is fully implemented, data will be integrated using both horizontal and vertical federation. Horizontal federation integrates data across dissimilar data sources. Vertical federation integrates different time slices or aggregations of the same data stored in different DWs as described above, essentially performing a union across the data sources. Vertical federation could be used, for example, to integrate both daily and monthly versions of the same data, or to combine trade data from two different time periods. Werr makes the point that within NYSE Euronext, vertical federation is just as important as horizontal federation.

Consuming applications – These are commercial BI tools, custom reporting tools and query tools accessed by both internal and external users. By the end of 2012, the company expects to have 500 active BI users globally.

DATA VIRTUALIZATION INITIATIVES. NYSE Euronext is in the process of implementing its data virtualization architecture, taking a carefully planned, phased-in approach. The first initiative in this effort is to establish the data virtualization layer and connect it to all of the databases that provide source data for analysis and reporting. As stated earlier, this is the foundation on which the company will build all of its future data virtualization development.

About 80% of database systems have been configured for connectivity to the data virtualization layer. According to Werr, "These include all of our transactional systems as well as many of our reference data systems. Often the reference data systems cannot be accessed directly so we create a mirror database for reporting purposes so we don't impact the performance of the underlying data source."

Werr described several expected benefits of this initial data virtualization layer.

Eliminate the need for point-to-point connections and security – NYSE Euronext has a huge environment and a complex network and security is critical. Creating point-to-point connections between every client and every database is a daunting task with the potential for security issues. "For us, it is much more efficient and streamlined to provide all of the data access points from the data virtualization middle tier back to the data sources. This enables us to organize our network topology, better control security in terms of how the data is used and make sure we understand what data is leaving the data warehouse."

Provide a formal way to track usage – The data virtualization layer captures all data access activity in a local repository: every query processed, who ran it, how long it took, what data was accessed, etc. The benefit is the ability to analyze these metrics to improve the data architecture as it evolves and improve capacity planning to determine how much infrastructure is needed and how much effort to spend on optimizing access to data.

Provide a centralized location for troubleshooting problems – The organization can better manage and resolve problems because the layers involved are greatly simplified versus a point-to-point architecture.

 Although a key strength of data virtualization is its ability to leave source data in place and provide an integrated view of the data in a virtual middle tier, we believe there is significant value in using data virtualization even if the user is only accessing a single data source.

Leverage single-source views and federation – One point Werr made was the following. "Although a key strength of data virtualization is its ability to leave source data in place and provide an integrated view of the data in a virtual middle tier, we believe there is significant value in using data virtualization even if the user is only accessing a single data source." Werr predicts that the company will reap huge savings just from this level of data virtualization, given the benefits described above. "The additional benefits of using data virtualization for data federation will be icing on the cake."

Once the data virtualization layer is configured to connect to all of the data sources, the next steps will be to 1) integrate reference data systems and 2) implement a streamlined, standardized enterprise reporting system. Fast access to consolidated reference data is important from a regulatory/compliance and a research perspective, and for reporting to both internal users and external customers. Examples are identifying how much the firm made across all products and the firm's market share in terms of transactions by type of product. Today, it can take more than a week to collect data from multiple sources and consolidate it for reporting purposes.

Werr describes this as a data architecture challenge. "Reference data has always been a problem. We want to do this right in two areas: how to access all the data and how to make sure the data in all of our systems can be integrated. We have over 300 different ways to describe a member firm."

Werr's team is working with other internal groups to gather business requirements. The corporate strategy team has identified the top two business priorities: revenue by product and customer, and transactions/volume by product and customer. Coming up with a universal definition for a product and customer will be the initial project over the next six months. "We have to architect it properly, support cross-business models of customer and product, tie the reference data back to transaction systems and the NYSE Technologies business, put the data virtualization layer in the middle for data access and then put the BI layer on top for analytics and reporting. It will take us years to do this across all of our reference data."

CUSTOMER ACCESS TO MARKET DATA. Concurrent with developing the overall data virtualization architecture, the company elected to implement a specific data virtualization project because of its high visibility and high payback potential. This project replaces an existing system that supports web access to market data for external users, such as listing companies and member firms. The National Market Data System (NMDS) is the backbone for external customers who visit the NYSE Euronext web site to view reference data. Examples are listing companies that want to see their performance compared to others and the market for their stocks, and member firms that want to view their liquidity status. Werr's team is migrating NMDS to the data virtualization environment by building a new schema on top of the existing database and replacing data access routines that are embedded in Java code and expensive to maintain. This is a good example of moving legacy code out of the application stack and into the data virtualization stack. NMDS is about 70% in production with estimated completion by the end of 2011.

Werr anticipates "huge savings" from NMDS. The organization will be able to retire a legacy outsourced system that costs about $5 million annually to maintain with a new system that will cost one-tenth of that (less than $500K). Werr offered this additional perspective on the benefits: "External customers need fast access to this information. With data virtualization, we have the opportunity to engineer the application properly, use the appropriate technology stack, reduce cost and improve performance. These enhancements will result in increased value and good visibility for the firm."

The Implementation Process

The first decision involved what technology to use to integrate disparate data for analysis and reporting. NYSE Euronext opted to virtualize access across its data systems using common data services rather than consolidate the data on a single data warehouse platform, which can be expensive and affect timeliness, according to Werr.

The company then did an initial market evaluation and chose Cisco early on as the tool to pilot for the planned data virtualization layer. This decision was based on the vendor's maturity in the industry and the product's ability to satisfy over 80% of the evaluation checklist. Cisco strengths included an industry-leading optimizer, good caching, an intuitive administrative toolset, extensibility through web services, the ability to capture usage metrics in a repository and the fact that the Cisco code was itself a framework that could be extended to add new features.

During the evaluation process, Werr's team worked closely with Cisco to add an important missing feature: the ability to execute pass-through SQL as is. NYSE Euronext has many legacy applications with embedded SQL and wanted to pass the SQL through without having to incur the huge time-to-solution cost of going back into every system to reverse-engineer the embedded SQL into Cisco views. Support for pass-through SQL makes the Cisco suite much more flexible when integrating legacy data systems. Another valuable benefit of this effort is the ability to capture and analyze in Cisco all of the usage metrics for these legacy systems. Werr commented that it was "nice to see Cisco collaborate with us on ways to evolve the technology. We have identified many good use cases for Cisco that deal with complexity, disparate data sources and high data volumes. We are trying to figure out what other improvements we can suggest that will help us as well as other customers."

> *The initial implementation went live in early 2011 and is still evolving ... The approach has been to start small, build subject matter expertise and then champion data virtualization within the firm so other IT pockets take ownership.*

As a result of the evaluation, the firm purchased Cisco in mid-2010. The initial implementation went live in early 2011 and is still evolving, as described above. The approach has been to start small, build subject matter expertise and then champion data virtualization within the firm so other IT pockets take ownership. Werr doesn't want his team to become a bottleneck in the evolution of the firm's data virtualization architecture.

Werr estimates it will take another two years to fully implement the data virtualization plan. The company now requires that every new system fit into the data virtualization architecture, so the effort centers around migrating older systems to the new strategy and architecture. This often entails more cultural, organizational and logistical challenges than technology issues, especially where there are multiple owners of data.

ADVICE. Werr offered the following "best practices" advice based on his experiences.

Take a step-by-step approach to implementing data virtualization – Werr has seen data virtualization projects fail when organizations jump right in and choose data integration as the first use case without first fully understanding the underlying data. "There can be data quality issues across the various suite or perhaps the data just doesn't integrate well for some reason. While data virtualization is a good mechanism for bringing disparate data together, the data still has to be 'integratable.' You don't want to put the ability to succeed with data virtualization in the way of data quality." His advice is to evolve the implementation step by step. Introduce data virtualization as an abstraction of the data sources, then layer the BI applications on top and after that gradually implement the more advanced federation capabilities data virtualization offers. This is the approach NYSE Euronext is taking. "Given our volumes of data and business requirements, we have to be very cautious about implementing even basic data federation."

Use an experienced partner for data virtualization technology – Werr has experience developing a middle tier data access solution from scratch and knows how much time, effort and cost is involved. "Data virtualization is one of the most complex things to get right and that is why it is still evolving in many people's minds." The lesson learned is to adopt an industry standard approach and partner with the vendor as an arm of the IT organization to build out data virtualization capability.

Pay attention to performance – One reason the company brought in MPP systems was to mitigate the performance challenge of data virtualization. The data virtualization tier can rely on MPP systems for query performance on high-volume data. "MPP runs the query faster and doesn't flood the network with data. Putting a software layer on top of a fast database doesn't add much latency. In fact, it might improve performance with intelligent caching of reference data." Users are unpredictable when it comes to ad hoc analysis and reporting. When a query involves a federated join, the data virtualization layer is governing the query and if it isn't efficiently written, data can spill over into the middle tier. "This is why we aligned our MPP and data virtualization implementations."

Werr summed up his advice this way. "You need to know the strengths of the technology and the weaknesses of your data to make the best use of data virtualization."

Summary of Benefits and Return on Investment

BENEFITS. Werr provided a long list of benefits the organization will realize over the coming years from its data virtualization solution. We have already described many of these and Werr offered this high-level summary.

Reduced footprint of software deployed through reuse of common data services and infrastructure – One key is the ability to make a common piece of logic reusable by multiple consumers with no need to propagate changes into multiple applications or propagate the organization to make those changes. The other is to push common functions that have nothing to do with the business – how data is accessed and managed – out of individual applications and into a shared middle tier, enabling applications to focus on the core business processes. Other examples of complex functionality that can be streamlined and concealed from applications through a data virtualization layer include performance, high availability, backup and recovery and caching support. These all represent ways to make the overall system faster and more resilient to better meet SLAs and avoid duplicate development costs.

Enhanced agility through a flexible data delivery infrastructure – The data virtualization layer makes applications and data sources independent, enabling the company to make changes at either end without changing the other. One example is the ability to quickly migrate from one database platform to another without changing the application layer. An extension of this is the ability to run a new application using the latest technology and still support a legacy version in parallel with the option to eventually turn off the legacy system.

Ability to combine data virtualization with MPP to optimize performance – Combining the ability to intelligently distribute data across powerful MPP systems and the ability to bring the data together in a unified way through data virtualization enables NYSE Euronext to optimize the overall performance of its data delivery environment.

RETURN ON INVESTMENT. Werr states that all of the benefits will result in savings and return on investment. We have already described the savings anticipated from migrating NMDS to a data virtualization: over $4.5 million annually.

Summary of Critical Success Factors

Werr cited three primary factors that will be critical to the company's ultimate success with data virtualization.

Ability to migrate existing SQL without modification – "Pass-through SQL will accelerate our ability to unwind a lot of the point-to-point systems and redirect them to the data virtualization layer without having to retrofit all of our applications that contain embedded SQL," said Werr.

Ability to capture how applications use data – As described earlier, every read call will go through the common data access layer, giving the company knowledge about all system activity with the opportunity to proactively evolve and improve the data architecture and performance architecture.

Cisco partnership for data virtualization – Using the Cisco suite enables the company to centralize and provide universal access to federated data without having to reinvent a technology that works well.

Future Directions

NYSE Euronext has a clear roadmap in place for its data virtualization adoption and is well on its way to full implementation as described above.

Pfizer

Organization Background

Pfizer Inc. is a biopharmaceutical company that develops, manufactures and markets medicines for both humans and animals. As the world's largest drug manufacturer, Pfizer's principal products include Lipitor, Celebrex, Viagra and Norvasc among many other well-known drugs. Pfizer operates globally with 111,500 employees and a presence in over 100 countries. The company was founded in 1849 and is headquartered in New York City, New York.

Pfizer Worldwide Research & Development (WRD) is a diversified R&D organization that supports Pfizer's human health care business units, such as Primary Care, Specialty Care, Established Products, Emerging Markets, and Oncology. Pfizer WRD is the largest biopharmaceutical R&D organization in the world. Its 2009 budget of $7.8 billion represents over 15% of Pfizer's 2009 annual revenue of $50 billion.

Worldwide Pharmaceutical Sciences (PharmSci) is a group of scientists within WRD responsible for enabling what drugs Pfizer will bring to market. This group designs, synthesizes and manufactures all drugs that are part of clinical trials and toxicology testing within Pfizer. These include Phase I, II and III clinical trials (pre-approval) as well as Phase IIIB and IV studies (post-approval). An example of the latter is testing a drug for a different use than the one for which it has already been approved.

PharmSci also develops commercial processes for drugs (e.g., formulating the active pharmaceutical ingredients from a powder into a packaged tablet), supports manufacturing processes and manages all budgets and activities as drugs advance through development to launch. PharmSci then transfers control and responsibility for the drug to Pfizer Global Supply.

For this case study, we interviewed Dr. Michael C. Linhares, Ph.D. and Research Fellow. Linhares heads up the Business Information Systems (BIS) team within PharmSci. BIS is staffed by three scientists who are also information delivery specialists and domain experts. BIS is responsible for portfolio and resource management across all of PharmSci's projects. This involves designing, building and supporting systems that deliver data to executive teams to help them make decisions regarding how to allocate available resources – both people and dollars – across the overall portfolio of projects. PharmSci can be involved in over 100 projects annually.

BIS is part of PharmSci's business operations group. BIS partners with Business Technology (BT), the name Pfizer uses for its IT groups. BT provides and maintains servers in data centers, standard software and technical support as services shared throughout the Pfizer organization.

The Business Problem

A major challenge for PharmSci is the fact that it has a complex portfolio of projects that is constantly changing. According to Linhares, "Every week, something new comes up and we need to ensure that the right information is communicated to the right people. The people making decisions about resource allocation need easy and simple methods for obtaining that information.

One aspect of this is that some people learn the information first and they need to communicate it to others who are responsible for making decisions based on the information. This creates an information-sharing challenge." Linhares estimates that there are 80 to 100 information producers within PharmSci and over 1,000 information consumers.

> *Every week, something new comes up and we need to ensure that the right information is communicated to the right people. The people making decisions about resource allocation need easy and simple methods for obtaining that information.*

There is also a diverse set of information that needs to be put together to give the executives on the project team a full picture of the project portfolio – financial data, project data, people data, data about the pharmaceutical compounds themselves, and project planning and scheduling data. This data is created in and managed by different applications, e.g., project planning, financial tracking and resource management. Each application was developed by a different team, the data is stored in multiple sources managed by different technologies, and the applications don't talk to each other. This makes it very difficult to access summary information across all projects. Examples would be identifying how much money is being spent on all projects in the project management system, what the next milestones are and when each will be met, and who is working on each project. "We needed a solution that would allow us to pull all this information together in an agile way."

When Linhares joined PharmSci, there was very little in the way of effective information integration. Most integration was done manually by exporting data from various systems into Excel spreadsheets and then either combining spreadsheets or taking the spreadsheet data and moving it into Access or SQL Server databases. "The process was very spreadsheet centric and involved a lot of manual steps," said Linhares. There were other issues with this "spreadmart" approach: spreadsheet data was separated from its source with no real security controls, the system lacked scalability and opportunities for reuse, multiple copies of the spreadsheets (with various changes) were available, and it often took weeks to build a spreadsheet with only a 50% chance that it would include all of the data required.

To be successful, the solution to these data integration and reporting problems had to provide the following:

• A single, integrated view of all data sources with a common set of naming conventions.

• A flexible data federation/virtualization layer that would be independent of both the data sources on the back end and the reporting tools on the front end to facilitate easy change management. For example, a change on one end (e.g., moving data from one type of database to another) would not require a change on the other. PharmSci also wanted the data virtualization layer to handle metadata and business rule functionality so there would be a single point for managing and monitoring the solution.

• A development suite that supported fast, iterative development and, therefore, continuous process improvement. As Linhares put it, "We often don't get it right the first time." This also would make it easier to justify developing "one-off" solutions that might be discarded after one-time use.

Linhares also pointed out that PharmSci was not constrained by the need to meet external regulatory requirements or good manufacturing practices. Examples of the former are FDA and SEC rules, which require systems to be validated with complex documentation. This gave BIS more flexibility in designing its solution.

BIS considered three solution architectures. The first was a traditional approach of an integrated, scalable information factory. Pfizer had already implemented information factories in WRD using a combination of Informatica ETL tools, Oracle databases and custom-built reporting applications. However, according to Linhares, an information factory "seemed like overkill. We didn't have high volumes of data, nor did we need the inherent complexity of using ETL tools to transform and move data while making sure we included all the detailed data we might possibly need over time." Furthermore, because of the way that information factories were managed within Pfizer, change management entailed significant overhead. However, the architectural concepts of an information factory were not going to be ignored in the final solution.

A second possible approach was to implement the solution in a single integrated technology (SQL Server with integration services). Major disadvantages were the lack of access to multiple data source types, the need to move data multiple times and the lack of an integrated metadata repository for understanding and organizing the data model.

The third option was to create a federated data virtualization layer that integrated and accessed the underlying data sources through virtual views of the data. By leaving the source data in place, this approach would eliminate the issues inherent in copying and moving all the data (which Linhares described as unnecessary, "non-value added" activities). With the right technology and mix of products, data virtualization would enable PharmSci to migrate from inefficient, off-line spreadmarts to online access to integrated information that could be rapidly tailored and reused to dramatically increase its value to the organization.

The Data Virtualization Solution

Pfizer's solution is the PharmSci Portfolio Database (PSPD), a federated data delivery framework implemented with the Cisco Data Virtualization Suite. Data virtualization enables the integration of all PharmSci data sources into a single reporting schema of information that can be accessed by all front-end tools and users. The implementation and use of the different tools and techniques has evolved over the past three years to provide a hybrid solution that meets the needs of the customers.

ARCHITECTURE. The overall PSPD data delivery architecture includes the following components (see Figure 1):

Trusted data sources – There are many sources of data for PSPD; they are geographically dispersed and store data in a variety of formats. Here are some examples to illustrate the power of data virtualization to integrate a multivendor, heterogeneous data environment.

- Enterprise Project Management (EPM) is a SQL Server database of WRD's drug portfolio project plans. It includes detailed project schedules and milestones.

- The Global Information Factory (GIF) is an Oracle-based data warehouse of monthly finance data.

- OneSource, a database of corporate-level drug portfolio information, is a particularly interesting case. OneSource is itself a unified set of Cisco views across several different sources built by another group within Pfizer. PSPD incorporates a view of this data as one of its data sources, in effect leveraging and reusing Cisco federated views.

- Flat files are provided by the Finance Department on actual resource use.

- SharePoint lists are small SharePoint databases accessed using a web service.

There are other data sources as well, including custom-built systems. As Linhares pointed out, "It doesn't matter what data sources we have. With a Cisco-based virtual approach, we are not limited by the types of data we need to access."

Data virtualization layer – The Cisco Data Virtualization Suite forms the data abstraction layer that enables the solution to be independent of the data sources and front end tools. It provides access to all of the data sources and delivers the data through virtual views. These views effectively present the PharmSci Portfolio Database as subject-specific data marts. The Cisco metadata repository presents data lineage and business rules.

Consuming applications – The flexibility of the solution is demonstrated by the varied reporting applications that use the information in PSPD. Examples include the following.

- SAP Business Objects for ad hoc queries, standard reports and dashboards.

- Tibco Spotfire for analytics and access to data through standard presentation reports.

- Web services for parameterized queries.

- Data services to provide data for downstream applications.

- QuickViews (web pages built using DevExpress, a .NET toolkit) for access to live data.

SharePoint portal – Branded as "InfoSource," this team collaboration web portal is the front-end interface that provides integrated access to PSPD data for all PharmSci customers through the consuming applications described above.

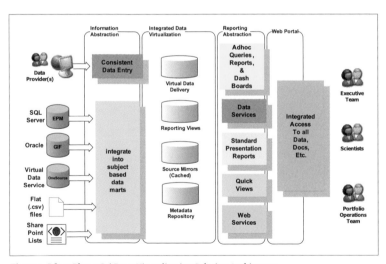

Figure 1. Pfizer PharmSci Data Virtualization Solution Architecture
Source: Pfizer

A COMPREHENSIVE DATA VIRTUALIZATION LAYER. Linhares stressed the importance of building two clear levels of abstraction into the data virtualization architecture. The first level abstracts sources (the information abstraction layer), the second consumers (the reporting abstraction layer).

"We started this process five years ago and it has made all the difference in the world. With a traditional data warehouse approach, an organization takes source data, moves everything to a staging area, then moves a portion of it again into the operational data store, and then moves some data a third time to a data mart. In the first stage of our federated implementation, instead of moving the data, we built a representation of the data in Cisco. If a source is ever changed by the owner, which often happens, we can update the representation in the information abstraction layer quickly. This allows control of all downstream data in one location. An example is migration of one of our sources from Oracle to Teradata. That transition can be very difficult in a traditional ETL/data warehouse environment. In our solution, all we have to do is point Cisco at Teradata instead of Oracle and adjust the view. With data virtualization, we don't have to be concerned about the technical solutions chosen by our data providers. Data virtualization enables a level of abstraction for us that separates us from the source systems and gives us automatic access to all of them."

The second level of abstraction is the one between the reporting schema and the front-end reporting tools. A consolidated and integrated set of information is exposed as a single schema across all of the source systems. Here, the solution allows BIS to be system agnostic and support the use of whatever tool is best for the customer. All of the reporting tools use the same reporting abstraction layer; they always get the same answer to the same question because there is only a single source of data.

BUSINESS RULES. Another key piece of the solution is the ability to include the business rules about how PharmSci manages its data within these abstraction layers. The business rules are embedded in the view definitions and are applied consistently at the same point. Linhares described two sets of business rules. One is applied in the representation of the source data and the other when requested data is sent out to the front-end tool/customer. On the data access side, for example, the employee data source contains over 900,000 possibilities who have worked or currently work at Pfizer, but PharmSci only needs information on active employees functioning in certain organizations. So the business rules reduce the retrieved data to the 90,000 people of interest. In the case of drug compound information, only compounds associated with active projects are made available from the source, reducing the data set from more than 10 million rows to a couple of thousand rows. An example on the reporting side is financial accounting period. Only projects with a red, green or yellow status are presented to the majority of users, although there are other possible status codes.

An important point here is that PharmSci has consolidated its business rules in a single location. Therefore, the Business Objects universe (the semantic metadata that describes the PSPD database to Business Objects) does not include any business rules. Embedding business rules in Business Objects and other front-end tools not only makes the tool environment complex, but also makes it more difficult to understand where the business rules are and to update them efficiently. "If we put all the rules in one place so we know where they are and how they are applied, it makes it easier to do change management and to explain how the data is being presented and delivered. It is also easier to understand interdependencies among the rules," said Linhares.

CACHING AND PERSISTENCE. The fact that Cisco has the ability to cache data from a virtual view as a file or to insert the data in a database table adds significant value to the solution. Caching can improve performance for iterative analysis by avoiding repetitive retrieval of the same data set. Local storage of data ensures 24x7 availability for access and analysis in cases where the source system is unreliable or down for maintenance reasons. "When we first started, no caching was used; however, it became clear very quickly that one key source was frequently unavailable." Local storage can also improve performance for certain data by eliminating network constraints/overhead. Other examples are when only a periodic snapshot of the data is required or when key data that is often requested does not change frequently.

With the ability to execute stored procedures on the federated views, Cisco essentially functions as a "light ETL" tool capable of moving small amounts of data from one database to another. Cisco saves the data to a database table using stored procedures scheduled with a trigger. "With data virtualization, we have the flexibility to decide what approach is optimal for us: to allow direct access to published views, which is purely virtual; to use caching to improve performance; or to use stored procedures to write to a separate database to further improve performance or ensure around the clock availability of the data."

ADDING NEW DATA SOURCES. The ability to easily add new data sources is another strength of the data virtualization approach. One example is the fact that BIS has developed web forms to capture data and store it in SQL Server. According to Linhares, "A lot of people were collecting data, such as financial forecasts, in Excel spreadsheets and we wanted to eliminate that. We now capture financial forecasts in a simple web page. The data is then combined, as part of PSPD, with portfolio and project information and actual financial information and reported out through Business Objects. It all integrates together very nicely with the help of Cisco."

Another more recent example is the capability to pull data out of SharePoint lists using data virtualization. The BT SOA team had to build a special web service application to do this, but Cisco now includes this functionality. Rather than rely on a manual process to export a spreadsheet and copy data into a database, there is now a web service in Cisco that accesses these lists as if the data were stored in a relational database.

A third example is the ease of sharing information already virtualized in Cisco with other Pfizer groups who also use Cisco. As described above, PSPD is already using a virtual view of the corporate drug portfolio as one of its trusted sources. Sharing PSPD information in return simply requires building a view and giving another organization access to it. As Linhares put it, "It is so easy to reuse and cross-share information across the organization as long as everyone is using the same environment."

APPLICATION DEVELOPMENT. Prior to data virtualization, the rate-limiting step (i.e., the slowest step) for BIS in fulfilling a customer request for information was data integration. Now it is designing an application that can effectively leverage data readily available in or easily added to the data virtualization layer. BIS uses a simple process for this. The first step is what Linhares calls "triage" – looking at what the customer wants, estimating how long it will take and communicating that to the customer. BIS does not spend a lot of time documenting the requirements of the solution. Instead, the group first creates a prototype on paper in the form of a simple data flow, then creates the necessary virtual views, gives the customer web access to the views and asks: "Is this what you wanted?" The customer can then play with the result and respond with any changes or additions needed. BIS arrives at the final solution working with the customer in an iterative process.

Linhares gave this example. "Let's say someone comes to us and asks for a dashboard with these ten items on it. Some or all of the requested information is new and not readily present in our existing data delivery system. Because of the way we have built our virtualization layer, it is relatively simple for us to obtain the data – it is most likely already sourced – and deliver the data in a short period of time. It usually takes us only one to three days to put together the data someone wants. The most difficult part of the process is actually getting access to a new source of data, as the owner needs to be identified and access requested through formal channels. Historically, without data virtualization, this process typically involved a large number of people, including a DBA, and the need to write custom scripts."

The Implementation Process

Pfizer has used data virtualization within the organization since 2005. The PharmSci Portfolio Database project started in 2008. The first challenge was integrating the data. According to Linhares,

"It took three months to get a full understanding of the data and build the views that form the data virtualization layer. The process involved documenting at least 13 data sources, identifying common or primary keys and figuring out how to integrate the data. We [BIS] had an advantage because we already had a thorough working knowledge of the various data sources. How to translate and combine the data is 'the art of the science.' With a data warehouse or information factory, it may take up to a year to learn all the different ways you need to combine the data." A related issue was defining a set of standard terms that would be used and interpreted consistently across multiple systems.

Another challenge was establishing a culture of information sharing. Many people view information as power – that is, the more they know the more valuable they are to the organization – and may not be comfortable sharing that information. Yet sharing and transparency is critical to increasing the value of the information to the overall PharmSci (and Pfizer) organization. PharmSci now asks team leaders to publish their information on a monthly basis so everyone in the organization knows what each team is doing, how projects are going and how the development process is functioning. The team leaders actively publish their results and BIS makes the information available to the rest of the organization through Cisco and InfoSource.

Summary of Benefits and Return on Investment

BENEFITS. Linhares described several major benefits of the data virtualization solution.

The ability to provide integrated data in context – Data virtualization has enabled BIS to replace isolated silos of data with a data delivery solution that integrates different types and sources of data into a comprehensive package of value-added information. Instead of only the team leader and a core group of eight to ten people knowing about a project, the entire organization has access to relevant project information.

The independence of the virtualization layer – "This is one of the huge benefits of using data virtualization. It allows me to manage and monitor everything in one place and it makes change management easy for BIS and transparent to users."

> *Because this building block approach doesn't take a huge amount of time, it is not a problem if we don't get it right the first time. One great example is a request that came in Friday morning and was completed by that afternoon. The customer's response was an amazed, 'What do you mean you already have it done?'*

Fast, iterative development environment – The data delivery infrastructure already exists in the data virtualization layer (defined data sources, standard naming conventions, access methods, etc.) so when a request for information comes in, BIS can quickly put it together for the customer. Linhares added, "Because this building block approach doesn't take a huge amount of time, it is not a problem if we don't get it right the first time. One great example is a request that came in Friday morning and was completed by that afternoon. The customer's response was an amazed, 'What do you mean you already have it done?'"

Another plus is that if customers realize that requests can be handled quickly, they may be more open to asking for new and/or enhanced information. "Can you add this to my report?" "Sure, no problem." Often, the alternative perception is that it will take many months and dollars to deliver on a request. That is one reason the BIS triage process is so important. It gives the customer a realistic time frame in which to expect results.

Elimination of manual effort throughout PharmSci – According to Linhares, people initially resisted going away from their spreadsheets. But once there was a single source for the data and it was all available through InfoSource, there was a dramatic reduction in the need to have meetings to reconcile spreadsheet data among teams.

RETURN ON INVESTMENT. When asked about ROI, Linhares acknowledged that it is difficult to measure accurately. He gave an anecdotal example of a comment made by one team leader: "You have saved me so much time because I no longer have to hold multiple meetings multiple times a month to reconcile all these different sources for the financial forecast. I have one trusted source for it. Everyone has the same set of data." The solution enables people to be more efficient and productive. Another measure of success is that people within PharmSci continue to ask for more enhancements to make the system more useful to them and to solve new problems.

Summary of Critical Success Factors

The most important factor contributing to the success of PharmSci's PSPD is the agility and flexibility of the data virtualization architecture. It is easy is make changes. New sources can be added quickly and combined with other data to provide the customer with information in the appropriate context. Without data virtualization, this process would be quite challenging, requiring multiple tools, the need to export/move data, etc. The ability to rapidly prototype solutions and provide results in a timeframe relevant to the customer is also key.

Another success factor was recognition of the business need and designing the solution with this need as the focal point, rather than basing the solution on traditional database design techniques. Linhares described this approach as understanding the business model and how the solution should look from the business person's perspective. For PharmSci, the project is central to everything. So Linhares established projects as critical master data entities with all the appropriate quality and governance controls, and then used data virtualization to integrate all the related project facts.

Future Directions

Linhares mentioned two key efforts going forward. One is to identify and pursue new areas in PharmSci where a data virtualization solution can add business value. Examples are integration of scientific information generated in the laboratories to make it more widely available, and more effective sharing of master data using web services.

The second is to make it easier and simpler for customers to get personalized information from PSPD. Linhares' goal is to "put more capabilities in hands of my customers and make them self-sufficient. I want to provide them with applications that can sit on any device they have – a smartphone, a tablet, a laptop – and give them the flexibility to look at the information in context, move the information around to view it from a different perspective, analyze it, quickly change the presentation. But the application has to be simple and intuitive to use, with little or no training required. You open up your smartphone, tap on your project portfolio application, the app calls a web service to connect to Cisco and your data is downloaded to your phone. If you only care about specific data, that's what you get. Finding all projects in Specialty Care with a red status and budget issues takes only a couple of clicks or finger taps. The entire dataset is available at your fingertips."

Linhares described this as an "information-as-a-service" (IaaS) approach to data delivery and considers it particularly important for senior management. "Today, senior leaders still ask people to get information for them. Our data access tools aren't that simple to use yet."

The data virtualization layer is an effective foundation on which to build an IaaS extension to PSPD for two reasons. First, the data has to be effectively organized for delivery. Cisco already provides that through simplified views that integrate the data while hiding all of the underlying complexity from the customer. Second, the customer device has to be able to connect to the PharmSci intranet and request the data. Linhares stated that Cisco offers "tons of flexibility" here, including support for web services, REST, Java, ODBC, JDBC, "in effect, whatever we need."

Cisco has recently added a new discovery capability that Linhares has been beta testing over the past year. This enables the customer to discover and understand relationships between data sources faster and apply corrective measures if a source doesn't support the necessary relationships. It will help BIS understand and integrate new data sources more quickly in the future.

Qualcomm

Organization Background

Qualcomm Incorporated is a global leader in next-generation mobile technologies. The company manufactures chipsets, licenses technology and provides communications services worldwide, primarily to the telecommunications industry. More than 180 telecommunications equipment manufacturers license Qualcomm inventions worldwide. In 1989, Qualcomm introduced CDMA (Code Division Multiple Access), a technology for wireless and data products that provided an important foundation for the evolution of today's 3G (third generation) wireless technologies.

Founded in 1985, Qualcomm is headquartered in San Diego, California. The company has more than 17,000 employees worldwide and operates in 139 countries. Annual revenue in 2010 was more than $10 billion.

For this case study, we interviewed Mark Morgan, IT Manager in the Enterprise Architecture (EA) group within Qualcomm's IT organization. EA is responsible for providing strategic guidance, plans and roadmaps to align IT initiatives with business strategies across the enterprise. A key component of this is identifying, evaluating and deploying new solutions in the areas of information architecture, technologies and business processes. Goals are rapid delivery of leading-edge solutions and documentation of best practices to facilitate reuse and adoption throughout Qualcomm.

The Business Problem

In 2009, Enterprise Architecture identified data virtualization as a compelling concept that could address some major business challenges facing Qualcomm. Morgan described two issues in particular that drove the interest in data virtualization.

 We are constantly challenged to get things done faster to maintain our leadership position in the mobile technology market. This is a tremendous struggle for any organization, but it hits especially hard when the industry changes as fast as ours does.

One is the fact that Qualcomm operates in a rapidly changing business environment. "We are constantly challenged to get things done faster to maintain our leadership position in the mobile technology market. This is a tremendous struggle for any organization, but it hits especially hard when the industry changes as fast as ours does." Therefore, the potential for increased agility and speed of execution was a key component and benefit of a data virtualization approach and a critical success factor for Qualcomm. "Without data virtualization, a project that involved moving data around, including building the processes to do that and developing subscriber-publisher handoff agreements, might take months. In contrast, because data virtualization technology leaves source data in place, it would enable us to prototype and get feedback to customers much faster and get applications up and running in a matter of weeks."

The second challenge is effectively managing multiple terabytes of data and ensuring the integrity of the data. Prior to implementing data virtualization, Qualcomm was moving significant amounts of data among multiple systems, including data warehouses and data marts. In some cases, the same data might be stored in ten different systems. Monitoring and maintaining all these systems and keeping the data in synch was costly. It required not only a lot of additional disk space to store the data but also significant time and effort in manpower resources. When data got out of synch, the process of reconciliation and re-synching was very time-consuming. "We were just shuffling the data around and it wasn't giving us any benefit. But we had no other option at the time," explained Morgan. Data virtualization would enable Qualcomm to manage each dataset in only one place, simplifying operational maintenance while reducing the cost of both the disk storage and the resources needed to maintain the data.

"The technology looked very promising and we kept tabs on the industry. When we were ready to move ahead with an evaluation of the technology, Cisco had stepped out as a leader in the data virtualization space and our proof-of-concept [POC] testing showed that Cisco could fulfill all of Qualcomm's data virtualization requirements."

The Data Virtualization Solution

Qualcomm's solution is an enterprise-wide data virtualization layer built with Cisco's Data Virtualization Suite. EA started the Cisco implementation with its POC test applications, which are described in more detail in the section on "The Implementation Process" below. After successful completion of the POC, the repository and virtual views created for the POC served as the foundation on which to expand the scope of the data virtualization environment across the enterprise.

The primary use case for data virtualization is the ability to make data available to multiple applications without having to copy and move the data. For example, when Qualcomm brings in a new application for a specific business purpose, the application may need access to data that resides in, say, five different repositories on five different systems – a data warehouse (DW), corporate directories, other applications, etc. Instead of spending six months on an integration project to bring the data physically together in one place to make the application functional, Morgan's group can build on the existing repository and virtual views within Cisco and simply plug the application into the data virtualization layer. If any required data is missing, "we can get access to that within a few days."

PEOPLE. "People" was one of the first applications implemented using data virtualization. "People" is Qualcomm's company directory, where anyone in the company can look up anyone else to find out the person's physical location, telephone extension, reporting hierarchy, etc. The application, which has been in existence for years, is very fast and the most heavily used web-based application in the company. "Originally, it involved a lot of application code written in Java to pull together the necessary data because the data was spread out across multiple data sources – LDAP [Light Data Access Protocol] directories, the human resources system, various legacy data stores and others. There were two problems with this. One, maintaining the app required significant time and effort. Two, the code to access the data was essentially 'throwaway' because it was not reusable. People who wanted to get at the same data for other reasons had to write their own code, just like we did. There were no APIs available to expose that data."

> *One big advantage, according to Morgan, is the ability to make changes much faster. "Before, adding a new data source meant writing new code and redeploying the application. Now, we simply point Cisco at the new data source. Cisco abstracts the data from the new source and the application doesn't know and doesn't care."*

With Cisco, Qualcomm was able to push the data access and integration functions down from the application layer to the data virtualization layer. One big advantage, according to Morgan, is the ability to make changes much faster. "Before, adding a new data source meant writing new code and redeploying the application. Now, we simply point Cisco at the new data source. Cisco abstracts the data from the new source and the application doesn't know and doesn't care. We also designed the architecture to be extensible, so when a new column is added to an existing data source, Cisco ensures that this new data shows up immediately." Having this level of independence between applications and data sources dramatically reduces the maintenance effort required to keep up with changing business requirements.

Another key advantage is the ability to easily create a web service in the data virtualization layer for accessing the underlying data sources. This provides a reusable web interface for anyone who wants access to the same data.

A third benefit of the People implementation was proving that Cisco could provide the required level of performance. As Qualcomm's most frequently hit web application – it is used hundreds of thousands of times every day – it was important that it be very, very fast. "One of the hallmarks of this application was that you could type in a name and get the information you wanted in milliseconds. You couldn't wait seconds. For us, it was a big test of Cisco to see if it could satisfy that level of real-time demand and maintain the same milliseconds response time. Cisco has done very well in this area."

OASIS. Another key data virtualization application was Oasis. Oasis was designed to help Qualcomm's product managers keep better track of Qualcomm's different chipsets by giving them better visibility into the overall design, development and manufacturing process – information about the cores, designs and manufacturers they depend on; where each chipset is in the design and development process; what new chipsets are being built; where in the manufacturing process each is; etc. Morgan described Oasis as "the ecosystem that provides our users with easy access to our chipset Plan of Record data."

This information was scattered across 10 to 15 different systems and project managers were expecting another 10 to 15 data sources to come online in the future. Without data virtualization, integrating all this data within the Oasis application would have taken years of effort, until 2015 or 2016. With data virtualization, it took less

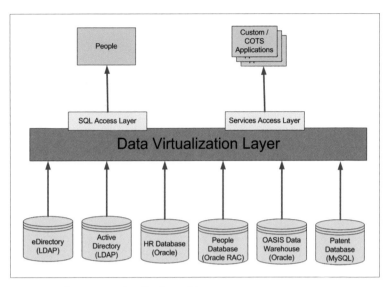

Figure 1. Qualcomm Data Virtualization Architecture
Source: Qualcomm

than 18 months. "We were tremendously successful because we were able leverage Cisco. There was already a data warehouse that integrated three or four of the data sources, so we started with that. We put Cisco on top of the data warehouse and the application on top of Cisco. So when new source systems came online, we simply added them to the Cisco data virtualization layer."

NEW IMPLEMENTATIONS. Another application, like Oasis, supports the chip manufacturing business and integrates three additional systems to further improve visibility into the manufacturing process. Qualcomm is a "fabless" company, which means that the company designs its own chips and then contracts out the fabrication ("fab"), or manufacturing, to specialized third-party chip manufacturers. Instead of building its own manufacturing facilities, Qualcomm uses fab capacity at partner companies. This is important given that the time lines for new chip design and production are very short and the company has to stay ahead of the industry.

This application provides real-time information about how the fabs are performing (e.g., the defect rates), which ones have availability and other information to ensure that manufacturing requirements are followed (e.g., chips manufactured in a particular location have to follow a certain path for testing, certification and assembly). It is a web application that delivers data to the same product managers as Oasis. Data sources include Oracle and Microsoft SQL Server and the volume of data is very large.

Implementation was challenging given the high performance requirements. Morgan stated that EA was able to get the application up and running quickly, but performance was too slow. "You would load a page and might have to wait 30 to 40 seconds to see it. Then you would click on something and it took another 12 to 15 seconds to display." Qualcomm brought Cisco on site to help with performance tuning. It turned out that the problems involved both query optimization and data architecture. One solution was to push more query processing down to Oracle and SQL Server so they could each do what they do well, rather than pull all the data into the Cisco layer. Another was to ensure that data sets for joins were as small as possible and that join orders were optimal.

"When we started, we could only get it tuned to a certain point and the application team was concerned that we couldn't do any better than that. It was reassuring to learn from working with Cisco that further tuning was possible and we could meet our performance expectations. The response time is now down to less than a few seconds. We were able to prototype the application quickly and then go back and make it fast. We are very happy with the results."

Overall, Qualcomm has implemented 13 data virtualization projects (applications) and 28 common data access services that encompass over 160 different service operations within the data virtualization layer. Figure 1 shows Qualcomm's data virtualization architecture.

The Implementation Process

The first step in the implementation process was recognizing the potential for data virtualization to speed up application delivery and reduce the cost and complexity of data management. The second step was to identify the types of solutions that would benefit from a data virtualization approach, such as those that would otherwise involve ETL or messaging to copy and move data. The third step was selecting Cisco as the industry leader in this area for product evaluation. At that point, EA decided to conduct a proof-of-concept (POC) test of the Cisco Data Virtualization Suite to see if it could meet Qualcomm's solution requirements and deliver the expected benefits. The POC process is designed to take a promising technology from concept to purchase and implementation. As Morgan put it, "Because the promises of data virtualization sounded so magical and mystical, we needed the POC to prove out two things. First, would data virtualization really increase the speed of development and deployment? Second, would it work in our environment?"

EA developed four use cases for the POC:

- People/Oasis integration – Virtualize the People application and the Oasis project/role data and integrate Oasis with People without requiring any code changes.

- Finance dashboard data – Virtualize a financial data export (a comma-separated values file format) combined with Excel metadata and expose both as a SQL interface to Qlikview.

- Qualcomm Patent Portal (QPP) virtualization – Expose patent data via data virtualization. This project was chosen because the EA team had no previous knowledge of the source data and wanted to prove out the learning process using Cisco. Morgan described this as a "real-world test case of how quickly we can learn a system and integrate it from start to finish using Cisco."

- LDAP/Web services – Prove the ability to consume LDAP data and expose it as SOAP services.

EA set up scenarios describing how Qualcomm would be using Cisco in these areas, including performance requirements. The POC would test whether Cisco could handle these use cases and meet the desired performance and speed of development criteria.

The POC took place in late 2009. EA had Cisco come on site and gave the company two weeks to implement the four use cases. The two-week timeframe was based on the speed-up EA felt it should expect with Cisco. Traditionally, the implementation would have taken 12 to 16 weeks. Morgan said that even though there were a couple of issues during the POC that caused a delay of two to three days, "amazingly enough, Cisco was still able to successfully complete all of the POC work within our two-week window." Qualcomm then purchased Cisco in January 2010.

One interesting perspective Morgan shared from the POC was the following. "Normally, if I ask an engineer how long something will take and the engineer says a couple of hours, I immediately know that in reality it will take a couple of days. But when Cisco engineers say something will take a couple of hours, they really mean a couple of hours. In fact, the work is usually done within a few minutes and then testing and debugging takes the rest of the two hours. We found this compelling and kind of a shock. There is no need to pad time estimates on development and deployment with Cisco."

ADVICE. Based on Qualcomm's experience, Morgan offered the following advice to other organizations considering data virtualization.

Use data virtualization where appropriate – One lesson learned is that data virtualization is not a panacea and does not provide a solution to every problem. As Morgan stated, "Data virtualization works well to quickly implement data integration. However, when we have large data sets that need to be joined, or real-time processing concerns that prevent us from taking advantage of Cisco's caching features, performance with data virtualization may not be quite what we need. In some situations, we tried to use virtualization because we had the technology. While it did, in fact, help us develop and deliver feedback to the customer faster, virtualization wasn't fast enough and using caching created delays. While data virtualization solves 80% of use cases, sometimes it is worth it to do a full physical integration [using traditional ETL to move data to a DW] or put in a real-time stream with messaging to get live data. On the other hand, if you can use data virtualization in even half of your potential use cases, you are still far ahead and much more agile in those use cases."

Information governance is key – Another caution is the need to understand the importance of a clearly defined information governance model, including how to manage the data virtualization environment. In Qualcomm's case, Cisco is one shared environment enterprise-wide, but there are multiple business units using it and contributing source data. These sources all have different downtime schedules and upgrade schedules that have to be accommodated. In addition, the governance model needs to document who is responsible for both the shared infrastructure and shared services. In the case of services, this includes who owns and supports them, how will they be versioned and updated, etc. On the other hand, Morgan added, "one of the things we love about Cisco is that we are able to move so fast with it on the development side. So we found that a blended model worked for us. Our production environment is well-governed, but we have a separate development environment in which developers can move very quickly without a lot of red tape to slow them down."

Performance-tuning expertise is critical – A third point Morgan made is that while Qualcomm can get integration projects done in a week versus six to eight weeks, there is a need to then tune those environments to get the best performance possible. So a critical resource is an expert in Cisco/SQL who can do the performance tuning. Even with this additional resource requirement, Morgan stressed that Qualcomm has experienced a significant net reduction in resources required for integration projects due to the increased speed of execution using Cisco.

Summary of Benefits and Return on Investment

BENEFITS. Morgan cited three primary benefits of Qualcomm's data virtualization solution.

Agility and speed of execution – A key benefit is the ability to get feedback to customers faster. "Before we had Cisco, we would spend significant time identifying the requirements and then developing the application and the data integration piece. It could be three months before the customer saw anything. Only then would we find out that it wasn't quite what the customer wanted. That's a long cycle. With Cisco, we can model the application in a few days and show it to the customer. If it's not what the customer needs, we may only need another day to tweak it. This shorter cycle allows us to be iterative and to 'fail fast.' We can spend three days and fail quickly. Then we simply iterate and try again."

Reduced support costs – Using data virtualization has lowered Qualcomm's IT support costs by replacing the need to rely solely on traditional integration technologies. "When we use data virtualization, we can avoid spending time supporting ETL or EAI [enterprise application integration] projects, synchronizing data, etc."

More efficient data management – Data virtualization has improved the efficiency of Qualcomm's data management in two ways. One is the fact that the company no longer has to store, manage and synchronize the same data in multiple locations. This has also enabled Qualcomm to establish stronger ownership of data as part of its governance effort. "In the past, ownership of data was a big challenge. Once data is copied, who owns the data gets fuzzy. In a data virtualization environment, ownership of the data is well-defined because it stays with the data."

RETURN ON INVESTMENT. Before starting the data virtualization implementation, EA developed an estimated ROI based on the expected reduction in development costs for the first few projects. This estimate was over $2 million in savings (see Figure 2). Although EA has not measured actual ROI achieved, Morgan stated that the organization knows empirically that it has saved significant time and money overall. "In some cases, development hasn't gone as fast as we anticipated. So we need to apply our lessons learned and include the results of our experience in our upfront ROI expectations going forward. But

Projects	Oasis	People	Finance Dashboard	BI	EAI / ETL
Level of Effort with / without Composite (developer days)	800 / 2,000 days	20 / 120 days	360 / 720 days	320 / 640 days	350 / 600 days
Savings in developer days / percent	1200 days 60%	100 days 83%	360 days 50%	320 days 50%	250 days 41%
Value gains	$1,200,000	$100,000	$360,000	$279,000	$250,000
Total	$2,230,000				

Figure 2. *Estimated Return on Investment from Data Virtualization*
Source: Qualcomm

data virtualization has definitely been worth the effort and investment. Not only has it saved us money, but it has accelerated our ability to meet business demands."

Summary of Critical Success Factors

Several aspects of the Cisco implementation were critical to Qualcomm's success in achieving its objectives.

Top down support – According to Morgan, top management support was crucial because many people were skeptical of using a data virtualization approach versus traditional data integration techniques. "We heard over and over, 'Can it really perform? It sounds like magic.' But we had proven the technology through our POC and we had support from our CIO, who made data virtualization a strategic initiative. This is not a technology that would have been successful in our environment if we had implemented it in pockets here and there. Data virtualization was viewed as a radical departure from traditional techniques and many people were not quite ready to embrace a new way of doing things."

The ability to leverage existing skill sets – A "huge win" for Qualcomm was the fact that Cisco was SQL–based and did not introduce a new development language. This enabled existing developers with experience in traditional data integration techniques to be comfortable working on Cisco projects with very little training. Morgan described this as a "massive benefit to getting people on board quickly. The need to develop a new skill set would have been a tremendous barrier to entry."

Reliability and performance of Cisco – Another major contributor to success was the fact that "Cisco worked. While there are other data virtualization alternatives out there, Cisco offers a very mature, strong and stable operating environment. It doesn't crash." The People app is a good example. "Everyone uses it every day and it works well."

Assigning a centralized team to advocate for successful adoption – Absolutely critical to Qualcomm's success was having one development team dedicated to helping other teams succeed with the technology. Morgan's team played that role and "it enabled us to establish a knowledge base so as new teams came on board, we could partner with them."

Future Directions

Data virtualization has already proven that it can contribute significant value and Qualcomm continues to implement new projects to leverage its investment in the technology. One key enhancement Morgan would like to see is the integration of a more real-time memory cache within the Cisco suite. This would eliminate the need to use separate products for capabilities such as near-real-time replication.

"In addition, implementation of the new incremental caching capability [versus full cache refresh] will be of tremendous benefit for applications that need real-time data, and will enable us to adapt Cisco to a wider variety of use cases and leverage it even more in the future." Morgan views this concept of a "trickle feed virtual layer" as the next step in the evolution of data virtualization.

Qualcomm also continues to improve and mature its information governance effort. "As the use of data virtualization expands, it exposes gaps in our existing model that we can then address to make our governance process more complete and comprehensive."

Conclusion

The Business Agility and Data Virtualization Journey

"Now this is not the end. It is not even the beginning of the end. But it is, perhaps, the end of the beginning."

In this profound statement, Prime Minister Winston Churchill characterizes the ongoing perseverance required of the British people in their struggles during World War II. Similarly, enterprises are challenged to continuously adapt and compete within their dynamic business environments. Business agility is critical.

Data virtualization is a data integration approach and technology that delivers needed business agility across a range of business challenges, functional domains and industries. This is clearly shown in the ten case studies included in this book. By significantly improving business decision agility, time-to-solution agility and resource agility, data virtualization provides enterprises with business agility value far beyond that which could have been achieved through traditional data integration methods alone.

However, it is important to understand that achieving business agility through data virtualization is a journey, not a one-time event. Others already on this journey have achieved significant business agility gains by starting small with focused solutions in advance of enterprise-scale expansion. To guide them on the technology side, they have found that a strong and ongoing partnership with the data virtualization vendor is key. And to optimize the development of skills, expertise and best practices, they have found that centralized ownership of the data virtualization effort is the best way to ensure success.

Go Forward with Confidence

As demonstrated in the case studies, other enterprises have blazed the trail and provided the business justification and technical roadmap required for effective adoption of data virtualization. Use their experiences, insights and lessons learned wisely.

If your organization is new to data virtualization, this book demonstrates with real-world examples how you can move beyond traditional data integration and use data virtualization to improve your organization's business agility.

If your organization is already adopting data virtualization, this book will help you successfully accelerate and expand your adoption, compound your business agility gains and achieve additional business and IT benefits from data virtualization.

We urge you to go forward with confidence.

Tangible Next Steps

Reading this book is a great "end of the beginning" in your business agility and data virtualization journey. Next steps include gathering additional data to successfully implement initial data virtualization-based data integration solutions and then expand your adoption.

GATHER MORE INFORMATION. Gathering more data virtualization information and insight can be a valuable next step. While this is the first book on data virtualization, others will surely follow. Analyst firms will continue to expand their research and advice on the topic, and leading BI media will continue to cover end-user adoption and vendor technology advancements. Further, professional associations such as TDWI and DAMA will continue to enhance their data virtualization education programs and certifications.

And don't discount the data virtualization technology vendors just because they want to sell you their data virtualization solutions. These companies have significant data virtualization domain knowledge spanning not only their products and services, but also all of the relevant people and process activities critical to successful data virtualization adoption. The Data Virtualization Leadership Blog and Data Virtualization Channel on YouTube are just two examples of the rich, multi-faceted information available from one of the leading data virtualization technology vendors, CIsco.

IMPLEMENT DATA VIRTUALIZATION. The ultimate goal of more complete, high-quality and actionable information is to make the right business decision and quickly move on to successfully adapt the business. Only then will the business realize the benefit of the change.

The same can be said for data virtualization. Its value comes from successful implementation and ongoing adoption. As seen in the data virtualization adoption patterns described earlier or in any of the case studies, there are many places to start. But start nonetheless.

EXPAND ADOPTION. Heed the lessons from Darwin and continuously adapt how and where you apply data virtualization. Proactively seek new opportunities. Develop expertise and evolve best practices. And keep learning more. Then use your successes and realized benefits as the foundation on which to further enhance your business agility.

Good luck in your journey.